The elements of nuclear power

Third edition

The elements of nuclear power

D. J. Bennet

Senior Lecturer in Nuclear and Mechanical Engineering
University of Strathclyde, Glasgow

J. R. Thomson

Central Systems Engineering Department
National Nuclear Corporation Limited

Longman Scientific & Technical
Copublished in the United States with
John Wiley & Sons, Inc., New York

Longman Scientific & Technical,
Longman Group UK Limited,
Longman House, Burnt Mill, Harlow,
Essex CM20 2JE, England
and Associated Companies throughout the world.

Copublished in the United States with
John Wiley & Sons, Inc., 605 Third Avenue, New York, NY 10158

First published 1972

Second edition 1981

Third edition 1989

British Library Cataloguing in Publication Data
Bennet, Donald J. (Donald John), *1928–*
 The elements of nuclear power.—3rd ed.
 1. Nuclear power
 I. Title   II. Thomson, J.R. (James Robert),
 1955–
 621.48

ISBN 0-582-02224-X

Library of Congress Cataloging-in-Publication Data
Bennet, D. J. (Donald John)
 The elements of nuclear power / D.J. Bennet, J.R. Thomson. — 3rd
ed.

   p.   cm.
 Bibliography: p.
 Includes index.
 ISBN 0-470-21317-5 (U.S.)
 1. Nuclear reactors. 2. Nuclear power plants.   I. Thomson, J.
R., 1955–   . II. Title.
TK9202.B43 1989
621.48'3--dc19   88-26666
                                                 CIP

Set in Plantin 10/12

Produced by Longman Group (FE) Limited
Printed in Hong Kong

BST
OPN

Contents

Preface to the third edition

In the sixteen years since the publication of the First Edition of this book, nuclear power has expanded greatly from the relatively few countries in which early developments took place to a large number of countries world-wide. Some of these countries now depend heavily on nuclear power for their energy supplies.

The purpose behind this Third Edition remains the same, to provide an introduction to the theory and technology of nuclear power which is suitable for students of mechanical and electrical engineering, and for engineers working in the field of nuclear power.

The serious reactor accidents at Three Mile Island in 1979 and Chernobyl in 1986, as well as the much publicized concern in the UK over radioactive effluent levels and leukaemia clusters have increased public awareness of the hazards of nuclear power and all aspects of the nuclear fuel cycle. The additional new material in this edition deals with some of these topics, in particular the effects of nuclear radiation on humans, the safety of nuclear reactors and those parts of the nuclear fuel cycle which deal with fuel element manufacture and the reprocessing of irradiated fuel.

The opinions expressed in this book do not necessarily reflect the views of the National Nuclear Corporation.

University of Strathclyde,   D. J. BENNET
Glasgow.

National Nuclear Corporation,   J. R. THOMSON
Cheshire.

June 1988

Preface to the second edition

In the eight years since the First Edition was published the use of nuclear energy for electricity generation has expanded throughout the industrialized world. This expansion seems set to continue, although in the present economic situation the demand for energy is growing slowly, and the rate of nuclear power growth is also likely to be slow.

At the same time public awareness of nuclear power, and in particular its hazardous aspects, has increased considerably and several pressure groups exist throughout the world whose aims are to oppose the development of this form of energy.

Thus at the present time nuclear power appears to be at a crossroads and it is difficult to predict its future direction and rate of development with any certainty. One thing however is sure, that in the next few decades, until other sources of energy can be exploited, nuclear power has a vital role to play in world energy supply.

With this in mind, the Second Edition contains new material designed to give a better perspective to the study of nuclear power — its history, its potential as an energy source, its benefits and its hazards. Anyone arguing the pros and cons of nuclear power should be aware of these aspects and it is hoped that this new material will be of help in this respect.

University of Strathclyde,   D. J. BENNET
Glasgow.

August 1980

Preface to the first edition

With the expanding use of nuclear fission as a source of energy for electricity generation, it seems likely that an increasing number of engineers and engineering students should be familiar with the principles underlying the generation of power from this source.

This book is based on a one-year course given to Final Year students of Mechanical Engineering at the University of Strathclyde. The aim of this course is to give these students an introduction to the principles of power generation from nuclear fission. The course begins with a description of the physical processes which take place in a nuclear reactor and then it develops simplified theory which enables calculations to be made for reactor criticality. Heat transfer in reactors, thermodynamic power cycles, reactor operation and radiation shielding are also dealt with briefly. A certain amount of additional material, notably the contents of Chapter 5, have been included in this book.

In writing this book I have derived much benefit from the co-operation during several years with my colleagues R. G. Tudhope and J. S. Lewis (now at Lanchester Polytechnic), and it is a pleasure to acknowledge my gratitude to them for the contribution they have made to this book by their discussion and opinions.

University of Strathclyde,   D. J. BENNET
Glasgow.

May 1972

Acknowledgements

We are indebted to the South of Scotland Electricity Board for permission to reproduce simplified illustrations of Hunterston 'A' and 'B' Stations.

List of symbols and subscripts

a	fuel element dimension
A	area, mass number
b	fuel element cladding thickness
B	breeding ratio, build-up factor
B^2	buckling
B_m^2	material buckling
B_g^2	geometric buckling
c	specific heat
C	concentration of delayed neutron precursors, source strength of gamma radiation, cost
d_e	effective diameter
D	dose rate
E	energy
E_A	available energy
f	thermal utilization factor, friction factor
F	neutron reaction rate per unit volume, feed
g	non-$\frac{1}{v}$ factor
h	enthalpy, heat transfer coefficient
H	rate of energy release per unit volume
I	effective resonance integral
$\mathcal{J}$	neutron current density
k	thermal conductivity, multiplication factor, Boltzmann's constant
k_∞	infinite multiplication factor
k_{eff}	effective multiplication factor
l	neutron lifetime
l_p	prompt neutron lifetime
L	length, height, diffusion length
L_s	slowing-down length
L'	extrapolated length
m	mass

$\dot{m}$ mass flow rate
M molecular weight, migration length
n neutron density
N density of atoms or nuclei, number of neutrons in nucleus
p pressure, resonance escape probability
P perimeter, product of enrichment
PP pumping power
q slowing down density, heat flux
q_L heat transfer per unit length of fuel element
Q rate of heat output or heat transfer
R radius, thermal rating, gas constant
R' extrapolated radius of core
R^+ extrapolated radius of reflector
s entropy
S surface area, source strength
t time
t_d diffusion time of thermal neutrons
t_m mean life of a radioactive isotope
$\bar{t}_m$ average mean life of delayed neutron precursors
T temperature, reflector thickness, reactor period, tails of enrichment
$T_{1/2}$ half-life of a radioactive isotope
T_d doubling time
u speed of neutron in the centre of mass system
U speed of nucleus in the centre of mass system
v neutron speed
V volume, value function for enrichment
W work, power
x isotopic fraction in enrichment process
Z atomic number
α temperature coefficient of reactivity, maximum fractional loss of energy after an elastic scattering collision
α_r radial form factor
α_z axial form factor
β delayed neutron fraction
γ fission product yield
δ reflector savings
δk excess reactivity
Δ mass defect
ϵ fast fission factor
η eta, efficiency
θ temperature difference, angle of scattering in the centre of mass system
κ inverse of the diffusion length
λ decay constant, mean free path

μ	linear absorption coefficient, viscosity
$\bar{\mu}$	average cosine of the angle of scattering in the laboratory system
ν	average number of neutrons produced per fission
ξ	logarithmic energy decrement
ρ	density, reactivity
σ	microscopic cross-section
Σ	macroscopic cross-section
τ	neutron age, shear stress
ϕ	neutron flux
ψ	angle of scattering in the laboratory system
χ	mass absorption coefficient
ω	reactor period

Subscripts

a	absorption
av	average
b	bulk
c	capture, core, coolant, centre of mass system
C	core (of reactor)
cl	cladding
CS	Compton scattering
D	dose
eq	equilibrium
E	energy
f	fission, film, feed, fuel surface
f_p	fission product decay
F	fuel, flux
G	gap
i	inelastic
l	laboratory system
M	moderator
MP	most probable neutron speed in a Maxwellian spectrum
p	constant pressure, product of enrichment
P	pump
PE	photoelectric
PP	pair production
R	reflector
s	scattering, surface, shutdown
sat	saturation
t	total, tails of enrichment
th	thermal
tr	transport
w	wall

γ gamma radiation
0 thermodynamic datum, initial conditions, origin of coordinates
1 fast group of neutrons
2 thermal group of neutrons

Modifying symbols

Where a symbol refers to a particular element, compound or isotope, this is indicated thus:

$N(H_2O)$ number of molecules of water per unit volume

$\sigma_c(^{238}U)$ the capture cross-section of ^{238}U

An historical introduction

The story of the discovery and development of nuclear energy, which in the context of this book is the energy released by the fission of uranium and possibly other heavy elements, may be taken as starting in 1932, the year in which Chadwick at the Cavendish Laboratory, Cambridge, identified the neutron.

This discovery was of great importance in several respects. Firstly, it enabled the structure of the atomic nucleus to be explained in a much more satisfactory way than had previously been possible, and it showed the possibility that any particular element might have a number of different isotopes, i.e. species in which the numbers of neutrons in the atomic nucleus may vary, one from another.

Secondly, the neutron provided atomic scientists with a new particle with which they could bombard atomic nuclei in order to induce artificial nuclear reactions. In previous years scientists had used high-energy protons and alpha particles (nuclei of the element helium) for this purpose, but soon after the discovery of the neutron it was realized by many of them, notably the Italian scientist Fermi working in Rome, that the neutron being uncharged (unlike the proton and alpha particle) would more readily penetrate the potential barrier of the atomic nucleus and interact with it.

In the following years Fermi and his colleagues in Rome bombarded many of the naturally occurring elements with neutrons and studied the products of the resulting reactions. In most cases he found that radioactive isotopes of the original elements were produced, and that when these isotopes decayed other elements, slightly heavier than the original elements, were produced. In this way uranium, the heaviest of the naturally occurring elements, was converted by neutron bombardment into heavier transuranium elements which do not exist naturally on the earth. Fermi made two other important discoveries at this time, namely that low energy neutrons are in general more effective than high energy neutrons for causing nuclear reactions, and that high energy neutrons can most effectively be slowed down to low energy by successive scattering collisions with light elements such as hydrogen in compounds such as water or paraffin wax. These two discoveries turned out to be of crucial importance in the development of nuclear energy in the following years.

Fermi's experiments with uranium were repeated by the German chemists Hahn and Strassmann who, in 1938, discovered that one of the products of the interaction between neutrons and uranium was barium, an element near the middle of the Periodic Table. Evidently a reaction had occurred in which the heavy uranium nucleus had, as a result of neutron bombardment, split into elements of intermediate mass. The physicists Meitner and Frisch, hearing of this discovery, evolved an explanation for the process on the basis of the liquid-drop model of the atomic nucleus, and calculated that an enormous amount of energy (far greater than in any previously known reaction) would be released as a result of the process, to which the name fission was given.

Other important features of fission were discovered in the following months. Joliot and his colleagues in France showed that some neutrons were emitted by the fission process, and it was later shown that these neutrons had very high energy. Thus the possibility existed that the fission process, initiated by a single neutron and producing two or three more neutrons, might be continued if these new neutrons caused further fission. The self-sustaining chain reaction so produced would be capable of releasing amounts of energy that were huge by existing standards.

Two distinct types of fission chain reaction were envisaged: one in which the process would proceed at a steady, controlled rate and release energy steadily and continuously; the other in which the fission rate would be so rapid and uncontrolled as to produce, literally, a nuclear explosion of considerable destructive potential. There were, however, many unknowns to be solved before these ideas even approached reality. Among these unknowns was the cross-section of uranium 235 for fission (the measure of the probability that this type of process would occur), and until this quantity was known there was no way of telling if a chain reaction would be possible, and if so how large would be the critical mass of uranium necessary. It was also realized that to achieve a chain reaction in certain types of system designed to give a steady and continuous energy release it would be necessary to reduce the energy of fission-produced neutrons to much lower energies at which, as Fermi had already shown, they would more readily cause further fission. The material to achieve this slowing down process became known as the moderator, and one of the earliest moderators to be used experimentally was heavy water, which at the time in question had only one source in Europe—the Hydro-Electric Company of Norway, from whom the French obtained the entire stock in 1940.

The discovery of fission in 1938 and the further developments of 1939, coming as they did just before the outbreak of the Second World War, could hardly have occurred at a more crucial time in world history. Had Hitler appreciated fully the significance of this discovery and encouraged his scientists to develop it, there is every possibility that Germany would have been the first country to produce atomic weapons and world history in the last thirty or forty years would have been very different. Fortunately, from the British point of view, Hitler was unimpressed by the discoveries of his atomic scientists, many of whom were Jews and were fleeing to Britain and America, and fission research in Germany was pursued with limited resources and priority. Fission research in France came to an abrupt end in June 1940 and two leading French scientists, Halban and Kowarski, came to Britain with France's vital stocks of heavy water.

Thus in the summer of 1940 Britain, already involved in a single-handed struggle with Germany, became the focal point of fission research. There was in the country in that year an impressive collection of the world's leading scientists, many of them refugees from Europe. There was an extraordinary sense of urgency, for it was realized that the first country to produce an atomic bomb would almost certainly win the war, and no one knew exactly what the Germans were doing. In that year the scientists in Britain made impressive progress and showed theoretically that an atomic bomb with a devastating explosive power could be made from uranium 235, which would itself require to be separated from uranium 238 in natural uranium by a process such as gaseous diffusion.

There was, however, still a very long way to go to produce the pure uranium 235 in adequate quantities and manufacture bombs, and in wartime Britain these steps were considered to be impracticable in view of the industrial resources that would be required and the vulnerability of the plant to air attack. It was decided to transfer nearly all the research, development and production work to the United States of America, where work on fission was in progress, though it was not as advanced, nor was it being pursued with the same degree of urgency, as in Britain. America, however, had the necessary industrial resources and, even after the outbreak of war with Japan, was immune from air attack.

The principal aim of fission research in America in the early nineteen-forties was to produce atomic bombs, and once America was involved in the war the sense of urgency increased. Two methods were proposed for obtaining pure fissile material for bombs: first, pure uranium 235 would be obtained from natural uranium by isotopic separation, using the gaseous diffusion process and uranium hexafluoride as the gaseous compound that would diffuse through a series of membranes. Second, the transuranium isotope plutonium 239, which was known to be fissile, would be produced from uranium 238 by neutron bombardment in a controlled chain-reacting system using natural uranium. Fission in the uranium 235 fraction of the natural uranium would sustain the chain reaction, and surplus neutrons would convert some uranium 238 to plutonium 239 which could be separated from the uranium. To these ends the huge gaseous diffusion plant at Oak Ridge, Tennessee, went into production in 1943 to produce uranium 235, and in December 1942 the first controlled chain-reacting system went critical at Chicago under the direction of Fermi, who several years previously had left Italy for the United States. In the following two years bigger, more powerful reactors (as the controlled fission systems came to be called) were built, culminating in the huge plutonium producing reactors at Hanford in Washington.

By the summer of 1945 enough uranium 235 had been produced at Oak Ridge, and enough plutonium 239 at Hanford to make the first atomic bombs. One was tested at Alamogordo in New Mexico, two were dropped on Japan and brought the Second World War to an abrupt end. The destructive power of these bombs confirmed the claims, and the fears, that had been expressed by scientists in the preceding years, and made August 1945 a watershed in human history.

In the years immediately following the Second World War the development of nuclear weapons proceeded apace, not only in America, but also in Britain and Russia who were determined to have their own weapons. Testing of atomic bombs, and the later and much more powerful hydrogen bomb, produced large-scale radioactive pollution in the atmosphere, and in due course these three countries signed an Atmospheric Test Ban Treaty to limit this pollution. The treaty was only partly effective, as France and China, new members of the 'atomic bomb club', did not sign the treaty and atmospheric testing continued, albeit on a reduced scale.

The early development of nuclear energy for electric power generation had its origins in military work. In the United States Admiral Rickover foresaw that naval vessels powered by nuclear reactors would have almost unlimited range, certainly much greater than the range of existing vessels, and this would confer considerable strategic advantages, particularly for submarines which would be able to cruise underwater for prolonged periods. It was, of course, necessary that the nuclear reactors should be as compact as possible to be installed in a ship, and this requirement led to the development of the first Pressurized Water Reactor, in which (as the name implies) water at high pressure is the moderator and coolant, and the fuel is slightly enriched uranium, i.e. uranium in which the uranium 235 content is increased by isotopic separation to 2 or 3 per cent. This type of reactor was much more compact than the huge graphite moderated Hanford reactors. The first nuclear powered ship, the submarine USS Nautilus, sailed in 1955. Two years later a reactor of similar design at Shippingport, Pennsylvania, became America's first commercial nuclear power plant for electricity supply, and in the thirty years since then this type of reactor, progressively enlarged in size and power, has been the basis for nuclear electricity generation in the United States and other countries.

Canada's wartime research had been centred on the development of reactors using heavy water as the moderator. This work was the continuation of the research started in France in 1939 and transferred to Britain between 1940 and 1942. Canada, being the Allies' sole producer of heavy water, was the logical country in which to continue this work when atomic research was transferred from Britain to the other side of the Atlantic. This early work has determined the course of Canadian nuclear energy development ever since, and present-day Canadian power reactors owe much to their original heavy water moderated reactors of the early post-war years.

In Britain at the end of the war the first priority was considered to be the production of plutonium 239 as a weapons material, and the first large British reactors were the two plutonium-producing reactors at Windscale. These reactors were similar to the American plutonium-producing reactors at Hanford in that they used natural uranium as the fuel and graphite as the moderator. This choice of materials was to some extent dictated by the fact that Britain did not at that time have access to large quantities of enriched uranium or heavy water, so the lines of development being followed in the United States and Canada were not then possible in Britain. The next stage was the construction of the combined power and plutonium-producing reactors at Calder Hall (beside Windscale) and when these reactors were commissioned in 1956 they became the world's first large power reactors to supply electricity to the public supply system. (A short time earlier the Russians had commissioned their first power reactor, but its output was very small.) In 1957 there was a serious fire in one of the original Windscale reactors, and both were closed down, however the Calder Hall reactors continued to operate successfully, and have become the prototypes for the subsequent development of gas-cooled, graphite moderated reactors for electricity generation in Britain, a development which has continued to the present day.

Such, then, is the brief history of the early development of nuclear energy in the western world. Similar developments of no less significance have taken place concurrently in Russia, and now many developed industrial countries, particularly western European countries and Japan, have their programmes of construction and use of nuclear power. The one type of reactor which more than any other now dominates the world nuclear energy scene is the American Pressurized Water Reactor which has been sold to, and built under licence, in many other countries.

There can be no doubt that the discoveries of the neutron, fission and plutonium have presented mankind with possibilities and dangers of an unprecedented magnitude. On the one hand there is the possibility of using for peaceful purposes an energy source which is potentially far greater than the world's resources of fossil fuel; on the other hand there are the dangers of the destructive power of nuclear weapons, a power that could destroy humanity. The discoveries of forty years ago cannot be undone or forgotten, and it may well be thought (bearing in mind that these discoveries and developments were made at a time when world war was in progress) that the birth pangs of nuclear energy might have been worse than they were. Two bombs were dropped in anger, and the destruction and death which they caused have given the world a lesson which so far has been heeded. The future use of nuclear energy presents challenges to the scientists and engineers who will be responsible for the design, construction and operation of nuclear reactors reliably and safely for power generation; it presents even greater challenges to mankind, and politicians in particular, to ensure that this source of energy is never again used for war.

Chapter 1

Atomic and nuclear physics—
a qualitative description

In this chapter we shall give a qualitative description of the structure of atoms and atomic nuclei, and of some related topics including nuclear reactions, mass-energy relationships, radioactivity and the properties of the radiation emitted by radioactive atoms.

The engineer or other person seeking some insight into the theory and operation of nuclear reactors is usually only concerned to know something about the results of atomic and nuclear reactions, and need not know the details or underlying theory of such reactions, which are often very complex. It is usually sufficient, therefore, for him to have a very simple picture of the atom and its nucleus which will be presented in this chapter. There are, of course, many textbooks of physics which give a detailed treatment of atomic and nuclear physics.

1.1 Atoms and nuclei

The atoms of all elements, which at one time were thought to be the fundamental particles of nature, consist of numbers of three more fundamental particles—protons, neutrons and electrons. The arrangement of these particles within the atom, and in particular the number of protons and electrons, determine the chemical identity of the element. The atom consists of a nucleus in which all the positively charged protons and uncharged neutrons are closely grouped together, and a number of negatively charged electrons moving in orbital paths around the nucleus. In an electrically neutral or unionized atom the number of protons is equal to the number of electrons, and this number, Z, is the atomic number of a particular element and identifies it. (This number corresponds to the position of the element in the Periodic Table.) The number of neutrons in the nucleus is denoted by N, and the sum of the number of neutrons and protons in the nucleus is called, for reasons that will shortly be apparent, the mass number, A.

$$N + Z = A$$

The term nucleon is applied to all particles, both protons and neutrons, in the nucleus.

Figure 1.1 is a useful, though not strictly accurate representation of an atom of carbon with six protons and six neutrons in the nucleus, and six orbital electrons. To be more accurate, the radius of the innermost electron orbit should be about ten thousand times the radius of the nucleus.

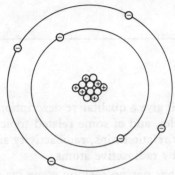

Figure 1.1. Atomic structure of carbon 12

1.2 Isotopes

Atoms having the same atomic number Z, but different numbers of neutrons N are called isotopes of the element identified by Z, and all elements have a number of isotopes, in some cases twenty or more. The naturally occurring elements each have one or more stable isotopes which exist naturally, and other isotopes which are unstable or radioactive and can be produced by artificial means. Different isotopes of an element behave identically as far as their chemistry is concerned, which is not surprising as chemical bonds exist between electrons. Isotopes differ from one another physically in that the masses and other characteristics of their nuclei are different, which is to be expected as it is in the nuclei that the difference between two isotopes lies.

The complete identification of an isotope is made by giving its chemical symbol, the atomic number Z as a subscript and the mass number A as a superscript. For example the symbol $^{16}_{8}O$ identifies the isotope of oxygen which has eight protons and eight neutrons in its nucleus. The isotope $^{17}_{8}O$ has eight protons and nine neutrons in its nucleus. Naturally occurring oxygen consists of a mixture of three isotopes, $^{16}_{8}O$, $^{17}_{8}O$ and $^{18}_{8}O$. There are also three radioactive isotopes of oxygen which do not occur naturally. The subscript Z is in fact unnecessary as the name oxygen identifies that element with eight protons in its nucleus and the symbols may be written as ^{16}O, ^{17}O and ^{18}O.

Hydrogen is an important element in nuclear engineering. Naturally occurring hydrogen consists of two isotopes, 99·985 per cent of the isotope ^{1}H and 0·015 per cent of the isotope ^{2}H called heavy hydrogen or deuterium. There is a third isotope ^{3}H called tritium which is radio-active. This is the only case in which the different isotopes of an element have different names; usually they are identified by their mass numbers.

1.3   The units of nuclear physics

The properties of protons, neutrons and electrons, and in particular their mass and charge, are important as they determine the way in which these particles behave. This is therefore a convenient point to introduce some of the units in common use in atomic and nuclear physics and nuclear engineering.

The unit of mass is the unified atomic mass unit (u). It is defined as one twelfth of the mass of a neutral carbon 12 atom. Its value is

$$1 \text{ u } = 1.6604 \times 10^{-27} \text{ kg}$$
$$\text{or alternatively} \quad 1 \text{ kg} = 6.023 \times 10^{26} \text{ u}$$

The atomic mass of any isotope is equal to the mass of one atom of that isotope expressed in u. The atomic masses of a few isotopes are given in Table 1.2. The atomic mass of an element is the weighted mean of the atomic masses of the naturally occurring isotopes of that element.

The mol, or to be more precise if SI units are being used, the kilogram—mol of any isotope is that quantity whose mass expressed in kilograms is numerically equal to its atomic mass. From the definition of the unified atomic mass unit stated above, it is evident that the mass of one mol of any isotope may be expressed in u as $6.023 \times 10^{26} \times$ the atomic mass.

Since for any isotope,

$$\text{number of atoms per mol} = \frac{\text{Mass of one mol}}{\text{Mass of one atom}},$$

and the mass of one atom expressed in u is the atomic mass, it follows that the number of atoms per mol of any isotope is 6.023×10^{26}. This number, called Avogadro's Number, is an important physical constant, applicable to all isotopes and elements.

$$\text{Avogadro's Number} = 6.023 \times 10^{26} \text{ atoms/kg-mol}$$

The statements above can also be applied to compounds and molecules, the mol being defined as that quantity whose mass, expressed in kilograms, is numerically equal to the molecular mass. Thus one mol of molecular oxygen, O_2, has a mass of 32 kg and contains 6.023×10^{26} molecules.

The electronic charge of the proton and electron are equal in magnitude and opposite in sign. The magnitude of this charge, known as the electronic charge, is $1·602 \times 10^{-19}$ coulomb.

The unit of energy is the electron-volt (eV) or alternatively the mega electron-volt (MeV). The electron-volt is defined as the energy acquired by a particle of unit electronic charge as it passes through a potential difference of 1 volt.

$$1 \text{ eV} = 1.602 \times 10^{-19} \text{ joule}$$

Table 1.1 summarizes the masses and charges of protons, neutrons and electrons. From this information it is clear that nearly all the mass of an atom is concentrated in the nucleus, and it is also clear that the atomic mass is approximately equal to the mass number.

Table 1.1. The mass and charge of protons, neutrons and electrons

	Mass (u)	Electronic charge
Proton	1·007 277	+1
Neutron	1·008 665	0
Electron	0·000 55	−1

Table 1.2. Atomic masses (u) of some isotopes

Isotope		Mass	Isotope		Mass
Hydrogen	^{1}H	1·007 825	Carbon	^{12}C	12·000 00
	^{2}H	2·014 10		^{13}C	13·003 35
	^{3}H	3·016 05	Nitrogen	^{14}N	14·003 07
Helium	^{3}He	3·016 03		^{15}N	15·000 11
	^{4}He	4·002 60	Oxygen	^{16}O	15·994 91
	^{5}He	5·012 3		^{17}O	16·999 14
Lithium	^{5}Li	5·012 5		^{18}O	17·999 16
	^{6}Li	6·015 12	Uranium	^{234}U	234·040 9
	^{7}Li	7·016 00		^{235}U	235·043 9
Boron	^{9}B	9·013 33		^{238}U	238·050 8
	^{10}B	10·012 94			
	^{11}B	11·009 31			

1.4 Mass defect and binding energy

A study of Tables 1.1 and 1.2 shows that the mass of an atom is not equal to the sum of the masses of its constituent particles. For example the mass of the ^{16}O atom is obviously less than the sum of the masses of eight neutrons and eight hydrogen atoms. Somewhere in the process of building the atom from its constituent particles the classical principle of conservation of mass appears to have been violated, and the difference between the mass of an atom and the sum of the masses of its constituent particles is known as the mass defect.

The explanation is to be found in the principle of the equivalence of mass and energy in which Einstein stated that mass and energy are different forms of the same fundamental quantity. In many reactions there is an interchange of mass and energy so that, particularly on an atomic scale, the laws of conservation of mass and conservation of energy are not valid when applied separately to a reaction, and must be replaced by the law of conservation of mass plus energy. In any reaction in which mass changes, a decrease of mass is accompanied by the release of energy, and an increase of mass corresponds to the absorption of energy.

The equivalence between mass and energy is expressed by the famous equation:

$$E = mc^2$$

where c, the speed of light, is $2 \cdot 998 \times 10^8$ m/s.

In SI units this equation can be stated as:

A mass change of 1 kilogram during any reaction is accompanied by the release or absorption of 9×10^{16} joules of energy, or:

$$1 \text{ kilogram} \equiv 9 \times 10^{16} \text{ joules}$$

A more useful relationship between mass and energy on an atomic scale can be obtained:

$$1 \text{ u} \equiv \frac{9 \times 10^{16} \times 1 \cdot 66 \times 10^{-27}}{1 \cdot 60 \times 10^{-13}}$$

$$\equiv 931 \text{ MeV}$$

Returning now to the mass of an atom and its constituent particles, the discrepancy mentioned already is explained by the fact that the formation of an atomic nucleus from its constituents is a process which, if it could be carried out in practice, would result in a release of energy, and this energy is equivalent to the mass defect. Alternatively, the process of breaking a nucleus down into its constituent particles would produce an increase of mass and an equivalent amount of energy would be absorbed in the process. This energy is known as the binding energy of the nucleus.

The mass defect, $\varDelta$, for any nucleus can be calculated from:

$$\varDelta = ZM_p + NM_n - M \tag{1.1}$$

where M_p, M_n and M are the masses of the proton, neutron and the nucleus in question. A more convenient formula for the mass defect makes use of the masses of the hydrogen atom, M_h, and the neutral atom in question, M_A. If electron binding energies, which are very small compared with nuclear binding energies, are neglected, then:

$$M_A = M + ZM_e \quad \text{and} \quad M_h = M_p + M_e$$

where M_e is the mass of the electron. An alternative formula for the mass defect of a nucleus is, therefore:

$$\varDelta = ZM_h + NM_n - M_A \tag{1.2}$$

Equations (1.1) and (1.2) are not precisely equivalent, but in view of the fact that electron binding energies are very small they may be regarded as equivalent for all practical purposes.

The equivalence between binding energy, expressed in MeV, and the mass defect, expressed in u, is:

$$\text{Binding energy} = 931 \times \text{Mass defect}$$

and

$$\text{Binding energy per nucleon} = \frac{931 \times \text{Mass defect}}{\text{Mass number}}$$

Example. Calculate the mass defect and binding energy per nucleon of oxygen 16 using the mass values in Tables 1.1 and 1.2.

$$\Delta = 8 \times 1.007\ 825 + 8 \times 1.008\ 665 - 15.994\ 91$$
$$= 0.137\ 01\ \text{u}$$
$$\text{Binding energy} = 0.137\ 01 \times 931$$
$$= 127.6\ \text{MeV}$$
$$\text{Binding energy per nucleon} = \frac{127.6}{16} = 7.97\ \text{MeV}$$

A plot of binding energy per nucleon versus mass number for all stable isotopes displays a systematic variation which is shown by the curve of Figure 1.2. For light elements the binding energy per nucleon rises rapidly with increasing mass number from a value of 1.1 MeV for

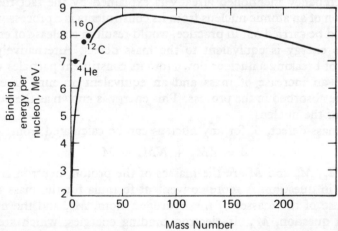

Figure 1.2. Curve of binding energy per nucleon versus mass number

^{2}H. A maximum value of about 8.8 MeV is reached for isotopes of mass number about 60, thereafter the binding energy per nucleon decreases slowly to a value of 7.6 MeV for ^{238}U.

A few isotopes lie well off the curve, three notable examples being ^{4}He, ^{12}C and ^{16}O. These are particularly stable or tightly bound nuclei which contain magic numbers of protons and neutrons. The magic numbers, 2, 6, 8, 14, 20, 28, 50, 82 and 126 correspond to the numbers of protons and neutrons that are required to fill shells of nucleons in the nucleus in much the same way that definite numbers of electrons are required to fill the electron shells of atoms. Nuclei containing magic numbers of protons or neutrons are usually very stable, and a nucleus containing a magic number of neutrons does not readily absorb another neutron, a point of some importance in nuclear engineering.

Study of the shape of the binding energy curve in Figure 1.2 gives some idea of the possibility of energy releasing nuclear reactions. It is clear from the shape of the curve that a reaction in which two very light nuclei can be made to join or fuse to form a heavier nucleus is a reaction proceeding in the direction of increased binding energy per nucleon. Similarly a reaction which causes a very heavy nucleus to split or undergo fission into two nuclei of intermediate mass proceeds in the direction of increased binding energy per nucleon. Any reaction which goes in the direction of increased binding energy, in other words a reaction whose products have greater binding energy than the original nuclei, is an energy-releasing or exothermic reaction since the binding energy represents energy released during the formation of the nucleus.

Thus the fusion of very light nuclei and the fission of very heavy nuclei are energy-producing reactions and are therefore of great interest to engineers in search of sources of energy. The problem is to create systems in which these reactions can take place under controlled conditions, and the methods of achieving a controlled fission reaction and extracting the energy released for the production of electrical power is the subject of this book.

The fusion of light nuclei are the thermonuclear reactions which are responsible for the enormous release of energy in the hydrogen bomb and the sun. A controlled thermonuclear reaction capable of providing the source of energy in a power station is one of the most challenging problems for physicists and engineers at the present, however when it is solved it will provide mankind with an unlimited source of energy.

An example of a fusion reaction is:

$$2\,^2H \rightarrow {}^3H + {}^1H$$

The total binding energy before the reaction is $2 \times 2\cdot23 = 4\cdot46$ MeV, and the binding energy of tritium is $8\cdot48$ MeV, so the energy release is $4\cdot02$ MeV.

1.5 Nuclear forces and energy levels

It is appropriate at this stage to mention very briefly the forces that exist between the nucleons in an atomic nucleus. The Coulomb electrostatic

force between charged particles is well known on a macroscopic
scale, and exists on an atomic scale between protons in the nucleus,
being a force of repulsion as they are all positively charged. The Coulomb force is therefore a force which tends to disrupt or burst the
atomic nucleus, and the fact that the nuclei of naturally occurring isotopes are stable and tightly bound indicates the existence of another
force which binds the nucleus together and is stronger than the Coulomb force. This is the case, and experiments have shown the existence
of a very powerful short range force of attraction that acts between
particles that are close to each other, within about 3×10^{-15} m. This
short range nuclear force acts with nearly equal strength between two
protons, two neutrons, or a proton and a neutron provided the separation is less than the distance quoted above, and it is this force which
binds the atomic nucleus together.

Normally atomic nuclei exist in an equilibrium or stable condition
known as their ground state of energy. However, as a result of nuclear
reactions (which might be caused by the bombardment of atoms by
protons, neutrons or other light particles), nuclei can be produced in
an excited or unstable condition in which one or a number of nucleons
are raised to an excited state. The excited states or levels in a nucleus
are similar to the excited states of atoms. In the case of the latter,
excitation results in an electron jumping from its normal orbit to
another orbit further from the nucleus, and an atom may have a number of discrete excited states corresponding to an electron having made
one or more such jumps. In the nucleus the situation is more complicated because excitation can result in several nucleons being raised to
excited levels simultaneously, and some nuclei can have a very large

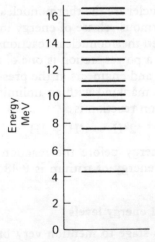

Figure 1.3. Energy levels of carbon 12

number of closely spaced excited levels. In general light nuclei have more widely spaced excited levels, and in all nuclei the spacing of the levels decreases as the excitation energy increases. Figure 1.3 shows schematically the energy levels of carbon 12.

Most excited nuclei exist in this state for only a very short time, a typical average lifetime being about 10^{-14} seconds, and they decay, or become de-excited, by the emission of high energy electromagnetic radiation known as gamma radiation, or particles such as neutrons, or both. In most reactions of interest to nuclear engineers involving the formation and decay of excited nuclei, the lifetime of the excited nucleus is so short that the process of formation and decay can be regarded as instantaneous.

1.6 Nuclear reactions

There are a great many possible nuclear reactions, of which only a few are of interest to us. These reactions are caused by the interaction of light particles such as neutrons, protons or deuterons (nuclei of deuterium), or gamma radiation with atomic nuclei. As an example we may take a reaction which is of importance in nuclear engineering resulting from the interaction between low energy neutrons and boron 10, which is written as follows:

$$^{10}B + {}^1n \rightarrow {}^7Li + {}^4He$$

As a result of this reaction lithium 7 and helium 4 are formed, the helium 4 nucleus is called an alpha particle. The reaction might be written in an abbreviated form as:

$$^{10}B(n, \alpha)\,^7Li$$

The lithium 7 may be formed at its ground state of energy, or at an excited state in which case it decays immediately to its ground state by the emission of gamma radiation.

Four fundamental laws govern all nuclear reactions:

1. *Conservation of nucleons.* The total number of nucleons before and after the reaction is the same.

2. *Conservation of charge.* The sum of the charges on all the particles before and after the reaction is the same.

3. *Conservation of momentum.* Since no external forces act during the reaction, the momentum of the particles is the same before and after the reaction.

4. *Conservation of mass plus energy.* Einstein's principle applies, and any loss of mass during a reaction is accompanied by a release of energy, or vice versa. The sum of mass plus energy before and after the reaction is constant.

To complete our example of a nuclear reaction we may apply these four laws to determine whether or not the $^{10}B(n, \alpha)\,^{7}Li$ reaction is exothermic, and if it is we can determine the energy of the lithium 7 and the helium 4 and the way it is shared between these nuclei. We will assume that the boron 10 atom is at rest before the reaction and that the kinetic energy of the neutron colliding with it is negligibly small.

It is quite clear that the reaction satisfies laws 1 and 2, there being eleven nucleons and a charge of five on each side of the reaction. Now comparing masses:

$$\text{Before the reaction:}\quad \begin{array}{ll} ^{10}B & 10{\cdot}012\ 94\ \text{u} \\ ^{1}n & \underline{1{\cdot}008\ 67\ \text{u}} \\ & 11{\cdot}021\ 61\ \text{u} \end{array}$$

$$\text{After the reaction:}\quad \begin{array}{ll} ^{7}Li & 7{\cdot}016\ 00\ \text{u} \\ ^{4}He & \underline{4{\cdot}002\ 60\ \text{u}} \\ & 11{\cdot}018\ 60\ \text{u} \end{array}$$

Since the total mass after the reaction is less than the total mass beforehand the reaction is exothermic and the release of energy is:

$$(11{\cdot}021\ 61 - 11{\cdot}018\ 60) \times 931 = 2{\cdot}80\ \text{MeV}$$

The energy of the system before the reaction was zero, so the energy after the reaction is 2·80 MeV, and this is shared as kinetic energy between the ^{7}Li and the ^{4}He nuclei, and in some cases as excitation energy of the ^{7}Li nucleus which in most reactions is formed at 0·48 MeV above its ground state.

It is clear from the assumption about the energy of the ^{10}B and the neutron that the momentum of the system is originally zero, and must therefore be zero after the reaction. The ^{7}Li and ^{4}He nuclei must therefore leave the site of the reaction in opposite directions in such a way that the magnitude of the momentum of each is the same. Applying these conditions to the case in which the ^{7}Li nucleus is formed at its ground state, and shares 2·80 MeV kinetic energy with the ^{4}He, it is easy to verify that the individual kinetic energies of the ^{7}Li and ^{4}He are 1·02 MeV and 1·78 MeV respectively. In the more likely event of the ^{7}Li being formed at 0·48 MeV above its ground state, the total kinetic energy of the two nuclei after the reaction is 2·80 − 0·48 = 2·32 MeV, and the individual kinetic energies are 0·84 MeV for the ^{7}Li and 1·48 MeV for the ^{4}He.

In concluding this section it should be noted that the law of conservation of mass plus energy applies to chemical reactions as well as nuclear reactions. For example, the reaction:

$$C + O_2 = CO_2$$

is a well-known exothermic reaction, so according to Einstein's principle the mass of a molecule of CO_2 is less than the mass of a molecule of oxygen and an atom of carbon. In this case, however, the mass difference is too small to be measured experimentally, and the energy release is much less than in nuclear reactions such as the $^{10}B(n, \alpha)\,^7Li$ reaction described above.

1.7 Radioactivity and radioactive isotopes

As has been stated already each of the elements has a number of isotopes. Some of these are stable and occur naturally, and others even at their ground state of energy are not stable and undergo spontaneous change known as radioactive decay. These are the radioactive isotopes which do not occur naturally (there are some exceptions which we shall meet later), but which can be produced artificially by nuclear reactions. For example, the interaction between neutrons and the stable isotope sodium 23 produces the unstable isotope sodium 24 by the reaction:

$$^{23}Na + {}^1n \rightarrow {}^{24}Na$$

As soon as the reaction starts and some sodium 24 is produced it starts to decay, and when the reaction is stopped it continues to decay until eventually there is none left.

If we plot the atomic number, Z, against the number of neutrons, N, for all the known isotopes it is seen that the stable or naturally occurring isotopes lie in a well-defined band shown in Figure 1.4. For the light elements a stable nucleus is achieved with approximately equal numbers of protons and neutrons, however for heavier elements it is necessary to have an excess of neutrons over protons for stability, and for the heaviest elements the number of neutrons is about one and a half times the number of protons. For example, lead 208 has 82 protons and 126 neutrons.

The upper limit of the stability region is bounded by $Z = 83$, the element bismuth. However, there are many naturally occurring radioactive isotopes up to $Z = 92$, uranium, whose existence will be explained later.

Radioactive isotopes lie for the most part outside the stability region of Figure 1.4, and when they decay the new isotope formed, called the daughter product, usually lies closer to the stability region than the original or parent isotope. If the daughter product is in the stability region it is likely to be stable, if it is outside the stability region and is radioactive it will decay in turn. In this way the original parent isotope may give rise to a chain of radioactive daughter products leading eventually to a stable isotope. Such a process can be represented by:

$$A \rightarrow B \rightarrow C \rightarrow D$$

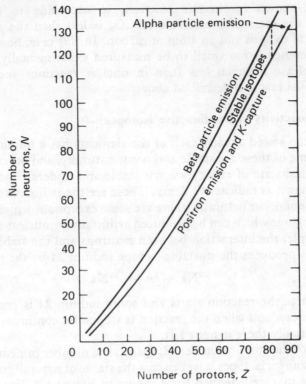

Figure 1.4. Plot of stable isotopes

in which the parent isotope A decays to form B which is radioactive and decays to form C, which in turn decays to form the stable isotope D.

There are four important types of radioactive decay.

1. *Emission of alpha particles.* As already mentioned an alpha particle is a nucleus of ^{4}He, a closely bound group of two protons and two neutrons. When a radioactive nucleus decays by alpha particle emission its Z number decreases by 2 and its A number by 4. Many of the heavy radioactive elements, both the naturally occurring ones with Z between 84 and 92, and the artificial transuranium elements with Z greater than 92 decay by alpha particle emission and as a result of the decay the daughter product is closer to the stability region than the original parent isotope.

Frequently the daughter nucleus is formed at an excited state of energy, i.e. the nucleons are not in a stable configuration. The excited nucleus decays immediately to its ground state of energy by the emission of gamma radiation, so the decay of a heavy radioactive isotope by alpha particle emission also produces gamma radiation. Uranium 238 is an example of a naturally occurring radioactive isotope which decays by alpha particle emission:

$$^{238}_{92}U \rightarrow {}^{234}_{90}Th + {}^{4}_{2}He + \gamma$$

2. *Emission of beta particles.* A beta particle is an electron emitted from an atomic nucleus. (The name was given to the particles before they were identified as electrons.) It may seem paradoxical that electrons can be emitted from an atomic nucleus which is composed of protons and neutrons, however this can be explained by assuming that within the nucleus a neutron is transformed into a proton and an electron, and it is this electron, called a beta particle, which is emitted.

Beta particles emitted by a radioactive isotope in this way do not all have the same energy, but have a spectrum of energies. The average beta particle energy is about one third of the maximum energy corresponding to the upper limit of this spectrum. In order to satisfy the principle of conservation of mass plus energy it has been postulated that another particle, called the neutrino (ν), is emitted with the beta particle. The existence of neutrinos has now been verified, however they interact very weakly with matter and are of no importance in nuclear engineering. The neutrino has zero charge, zero (or very small) mass and carries on average about two thirds of the maximum beta particle energy.

The transformation in the nucleus can be represented by:

$$^{1}_{0}n \rightarrow {}^{1}_{+1}p + {}^{0}_{-1}\beta + \nu$$

Beta particle decay has the effect therefore of transforming one of the neutrons in the nucleus into a proton, and the daughter nucleus has the same mass number as the parent, but its atomic number is greater by 1, as the following example shows:

$$^{60}_{27}Co \rightarrow {}^{60}_{28}Ni + {}^{0}_{-1}\beta + \nu + \gamma$$

As in the previous example the daughter product, nickel 60 in this case, is formed above its ground state of energy and decays by the emission of gamma radiation. Decay by beta particle emission occurs principally with isotopes having an excess of neutrons, i.e. isotopes lying to the left of the stability region in Figure 1.4.

3. *Positron emission.* Radioactive isotopes which have an excess of protons in the nucleus, i.e. those lying to the right of the stability region in Figure 1.4, may decay by positron emission. The positron is a particle whose mass is the same as that of the electron, but which is positively charged. It may be regarded as a positively charged beta particle which is formed in the atomic nucleus by the conversion of a proton to a neutron. As in the case of beta particle emission neutrinos

are emitted with the positrons. The process can be represented by:

$$_{+1}^{1}p \rightarrow _{0}^{1}n + _{+1}^{0}\beta + \nu$$

The positron is unstable and reacts with an electron to cause the annihilation of both particles and the production of gamma radiation. This is an example of a reaction in which mass is completely destroyed and an equivalent amount of energy is released in the form of electromagnetic radiation, and it is a striking example of mass–energy equivalence.

An example of positron emission is provided by the decay of iron 53:

$$_{26}^{53}\text{Fe} \rightarrow _{25}^{53}\text{Mn} + _{+1}^{0}\beta + \nu + \gamma$$

4. *K-capture*. Radioactive isotopes lying to the right of the stability region may also decay by a process known as K-capture in which an electron in the innermost or K-shell of an atom is captured by the nucleus and combines with a proton to form a neutron with the emission of a neutrino:

$$_{+1}^{1}p + _{-1}^{0}\beta \rightarrow _{0}^{1}n + \nu$$

An example of such a process is:

$$_{4}^{7}\text{Be} + _{-1}^{0}\beta \rightarrow _{3}^{7}\text{Li} + \nu + \gamma$$

Positron emission and K-capture processes have the same result, namely the atomic number of the parent atom decreases by 1 while the mass number remains the same. A radioactive isotope with an excess of protons thus moves towards the stability region as a result of decay by one of these processes.

When a daughter product is formed in an excited state and its nucleus decays to the ground state the gamma radiation emitted has an energy which is characteristic of the daughter product, and this radiation can also be used to identify the parent isotope by means of a detection system which not only detects gamma radiation but also measures its energy. As an example of this, the decay of caesium 137 by beta particle emission to form barium 137 may be considered. The barium is formed at an excited state and decays to its ground state by the emission of gamma radiation of 0·66 MeV energy. This radiation although it originates in the barium is taken as being characteristic of caesium 137 and the decay of this isotope may be measured by detecting the 0·66 MeV radiation.

The complete decay scheme of a radioactive isotope may be represented by Figure 1.5, however it should be remembered that sometimes, as in the case of the decay of strontium 90, the daughter product is formed at its ground state of energy and no gamma radiation is emitted.

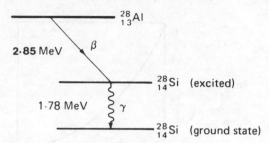

Figure 1.5. Decay scheme of a radioactive isotope

1.8 The rate of radioactive decay

Thus far we have described radioactive decay without any reference to the rate at which this process occurs. The decay of radioactive atoms is a random process, and the rate of decay can only be described statistically, however any sample of radioactive material likely to be of practical interest contains so many atoms that a statistical prediction about its behaviour turns out to be very accurate.

Different radioactive isotopes decay at different rates, but in every case the decay rate of a sample of a radioactive isotope at any time is directly proportional to the number of atoms of the isotope in the sample at that time. This statement can be written as

$$\frac{\mathrm{d}N}{\mathrm{d}t} = -\lambda N \tag{1.3}$$

where N is the number of atoms of the radioactive isotope in question at time t (the negative sign indicates that this number is decreasing with time), and λ is a constant, called the decay constant, which has a particular value for any radioactive isotope.

The concentration $N(t)$ at time t of a radioactive isotope may be determined by integrating equation (1.3), written in the form $\mathrm{d}N/N = -\lambda\mathrm{d}t$, to give

$$N(t) = N_0 e^{-\lambda t} \tag{1.4}$$

where N_0 is the concentration when $t = 0$. This variation is shown in Figure 1.6.

The unit of radioactivity which was in general use for many years until the introduction of SI units is the curie (Ci), defined as 3.7×10^{10} disintegrations per second. This unit was based on early measurements of the activity of one gram of radium 226. The curie has now been superseded by the SI unit, the becquerel (Bq), which is defined as one disintegration per second. This is a very small amount of radioactivity, and levels of radioactivity of interest are more likely to be expressed in kilobecquerels or megabecquerels.

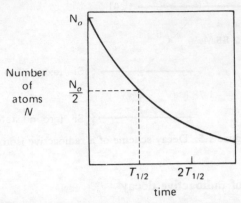

Figure 1.6. The decay of a radioactive isotope

A quantity that is more commonly used than the decay constant to characterize the rate of decay of a radioactive isotope is its half-life. The half-life is defined as the time for the number of radioactive atoms to be reduced to one half of the original value. Applying this definition to equation (1.4) we get:

$$\frac{N_0}{2} = N_0 \, e^{-\lambda T_{1/2}} \tag{1.5}$$

where $T_{1/2}$ is the half-life. The solution of this equation is:

$$\lambda T_{1/2} = \log_e 2 = 0.693$$

$$T_{1/2} = \frac{0.693}{\lambda} \tag{1.6}$$

It is not possible to attribute a precise lifetime to a radioactive isotope to express the time taken for it all to decay. However the fraction of the original quantity left at any time can be expressed since after n half-lives this quantity is $(\tfrac{1}{2})^n$. For example, after ten half-lives the fraction of the original quantity left is about one thousandth.

The mean life of a radioactive isotope is defined in the usual statistical manner, and the mean life t_m of N_0 radioactive atoms may be determined as follows:

$$t_m = \frac{1}{N_0} \sum_{\text{all atoms}} \left(\begin{array}{l} \text{Number of atoms which decay} \\ \text{in the time interval } t \text{ to } t + dt \end{array} \right) \times t$$

Transforming this to an integral:

$$t_m = \frac{1}{N_0} \int_{N_0}^{0} t \, dN$$

where $\mathrm{d}N$, the number of atoms decaying in the time interval t to $t + \mathrm{d}t$ can be shown from equations (1.4) and (1.5) to be $\lambda N_0 \, e^{-\lambda t} \, \mathrm{d}t$. Therefore

$$t_m = \lambda \int_0^\infty t \, e^{-\lambda t} \, \mathrm{d}t$$

$$= \frac{1}{\lambda} \tag{1.7}$$

The mean life is the reciprocal of the decay constant, and comparing equations (1.6) and (1.7) it is seen to be equal to $1 \cdot 44$ times the half-life.

1.9 Radioactive decay chains

Many practical problems involving radioactive isotopes are more complicated than the simple decay of a single isotope. In the event of the daughter product being radioactive it too will start to decay once it is formed, and there are many examples in nuclear engineering of a radioactive isotope being formed by the interaction between neutrons and a stable isotope in a nuclear reactor, and thus undergoing production and decay simultaneously. The analysis of problems such as the two just mentioned can be tackled by setting up and solving the appropriate set of differential equations which describe the rate of formation and decay of the isotopes involved, provided the initial concentrations at some time $t = 0$ are known. A hypothetical example will serve to illustrate the technique.

Example: Consider the radioactive decay chain:

$$A \rightarrow B \rightarrow C \, \text{(stable)}$$

Isotope A decays with decay constant λ_A to form isotope B which decays with decay constant λ_B to form the stable isotope C. At time $t = 0$ the concentration of A is N_{A0}, and the concentrations of B and C are zero. Determine the concentrations of A, B and C as functions of time.

From our earlier consideration of the decay of a single isotope we can use equation (1.4) to express the concentration of A:

$$N_A(t) = N_{A0} \, e^{-\lambda_A t} \tag{1.8}$$

For isotope B we may make the following statement:

$$\begin{pmatrix} \text{The rate of change of} \\ \text{the concentration of B} \end{pmatrix} = \begin{pmatrix} \text{The rate of formation} \\ \text{of B by the decay of A} \end{pmatrix} - \begin{pmatrix} \text{The rate of} \\ \text{decay of B} \end{pmatrix}$$

or:

$$\frac{\mathrm{d}N_B}{\mathrm{d}t} = \lambda_A N_A - \lambda_B N_B$$

Rearranging and using equation (1.8):

$$\frac{dN_B}{dt} + \lambda_B N_B = \lambda_A N_{A0} \, e^{-\lambda_A t}$$

Multiplying by the integrating factor, $e^{\lambda_B t}$, we get:

$$\frac{d}{dt}(N_B \, e^{\lambda_B t}) = \lambda_A N_{A0} \, e^{-(\lambda_A - \lambda_B)t}$$

Integrating:

$$N_B \, e^{\lambda_B t} = \frac{\lambda_A}{\lambda_B - \lambda_A} N_{A0} \, e^{-(\lambda_A - \lambda_B)t} + K$$

From the initial condition that $N_B = 0$ at $t = 0$:

$$K = \frac{\lambda_A}{\lambda_A - \lambda_B} N_{A0}$$

and the complete solution for N_B is:

$$N_B(t) = \frac{\lambda_A}{\lambda_B - \lambda_A} N_{A0} \, (e^{-\lambda_A t} - e^{-\lambda_B t}) \qquad (1.9)$$

N_C can now be found easily by noting that the total number of atoms must remain constant, i.e.

$$N_A(t) + N_B(t) + N_C(t) = N_{A0}$$

from which we obtain the result:

$$N_C(t) = N_{A0} \left(1 - \frac{\lambda_B}{\lambda_B - \lambda_A} e^{-\lambda_A t} + \frac{\lambda_A}{\lambda_B - \lambda_A} e^{-\lambda_B t} \right) \qquad (1.10)$$

Figure 1.7 shows the variation of N_A, N_B and N_C for the case in which the half-life of B is twice as long as the half-life of A.

1.10 The naturally occurring radioactive isotopes

There are three natural radioactive decay chains whose parent isotopes have very long half-lives. The three isotopes are uranium 238 ($T_{1/2} = 4.51 \times 10^9$ years), uranium 235 ($T_{1/2} = 7.1 \times 10^8$ years) and thorium 232 ($T_{1/2} = 1.41 \times 10^{10}$ years), and their decay chains contain many radioactive isotopes leading eventually in each case to a stable isotope of lead. The decay chain originating from uranium 238 is shown in Figure 1.8.

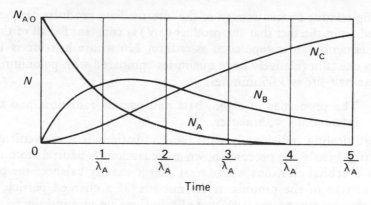

Figure 1.7. The concentration of isotopes in a decay chain

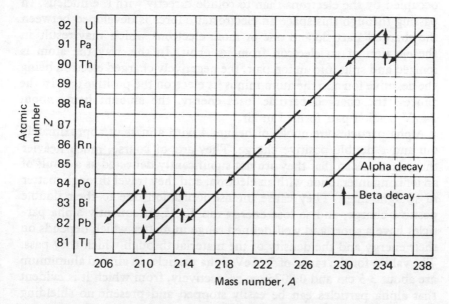

Figure 1.8. The natural radioactive decay chain of uranium 238

Many of the intermediate daughter products in these decay chains have quite short half-lives, but their existence is explained by the fact that they are being continuously produced. The half-lives of the parent isotopes of the three chains, such as uranium 238, are so long that during a period of a few years or even a few hundred years they can be regarded as decaying at a constant rate, and all their daughter products are in equilibrium, that is their rates of formation and decay are equal and their concentrations remain constant.

The relative concentrations of the intermediate products can be inferred from the fact that the product (λN) is constant for all of them. Consequently an isotope such as radium 226 whose half-life is 1600 years exists in relatively large quantities compared with polonium 218 whose half-life is 3·05 minutes.

1.11 The properties of alpha, beta and gamma radiation, and their interactions with matter

Before dealing individually with these radiations it is convenient to describe briefly the process known as ionization. A neutral atom contains Z orbital electrons whose total charge exactly balances the positive charge of the protons in the nucleus. If a charged particle approaches an atom it is much more likely (bearing in mind the relative radii of the nucleus and the electron orbits) to pass through the space occupied by the electrons than to collide directly with the nucleus. In passing through this space an electrostatic force is developed between the charged particle and one or more electrons which may result in the ejection of an electron from its orbit. In this way the atom is ionized and split into an ion pair, the negatively charged electron being the negative ion and the atom minus its electron the positive ion. In the process the charged particle loses energy, the amount being about 34 eV per ion pair formed in air.

Alpha particles are nuclei of helium 4 with a mass of approximately 4 u and a double positive charge. They are, of course, much heavier than electrons so that they are not significantly deflected as a result of an ionizing interaction with an electron, and they travel through matter in straight lines. They cause intense ionization due to their double positive charge, and so lose energy rapidly. Consequently alpha particles have a short and well-defined range in matter which depends on their energy and the density of the material through which they pass. The values for the range of 5 MeV alpha particles in air and aluminium are about 3·5 cm and 0·0025 cm respectively, from which it is evident that alpha particles can be easily stopped and present no shielding problem in nuclear physics or engineering, although they do present a health hazard if an alpha-emitting isotope is ingested in the body.

Unlike alpha particles which are emitted with discrete energies, beta particles are emitted from a radioactive isotope with a spectrum of energies. They interact with matter by causing ionization, but the effect is less intense than that due to alpha particles because of the smaller charge and much smaller mass of the beta particles. The range of beta particles emitted by a radioactive isotope is not well defined for two reasons, firstly the fact that they are produced with a spectrum of energies from zero up to their maximum energy, and secondly the fact that they do not travel in straight lines, but in zig-zag paths resulting from a change of direction at each ionization interaction. As a guide, however, the ranges of beta particles of 5 MeV energy in air and aluminium are about 2000 cm and 1 cm respectively, so that although they are much more penetrating than alpha particles they do not present a serious shielding problem; there is however a similar ingestion hazard.

Gamma radiation is electromagnetic radiation of very short wavelength (10^{-11} to 10^{-15} m) as compared with the visible spectrum (10^{-6} to 10^{-7} m). According to the quantum theory electromagnetic radiation can be regarded as being propagated as discrete quanta or photons whose energy, in joules, is given by the equation:

$$E = \frac{hc}{\lambda}$$

where h is Planck's constant, $6 \cdot 625 \times 10^{-34}$ joule seconds, c is the speed of light, metres/second, and λ is the wavelength of the electromagnetic radiation, metres.

This is an example of De Broglie's hypothesis, which has been well established by experiment, and which states that electromagnetic radiation has certain particle-like characteristics, and that particles such as electrons and neutrons exhibit certain wave characteristics such as diffraction.

There are three important methods of interaction between gamma radiation and matter—the photoelectric effect, Compton scattering and pair production. In the photoelectric effect the entire energy of the gamma photon is transferred to an orbital electron and the photon disappears. In the Compton scattering process a fraction of the energy of the photon is transferred to an orbital electron and a scattered photon emerges with reduced energy. In the pair production process a gamma photon of energy greater than $1 \cdot 02$ MeV is annihilated in the neighbourhood of a heavy nucleus and an electron-positron pair is created. This is an example of the conversion of energy into mass and is the converse of the positron–electron annihilation described in connection with radioactive decay.

Gamma radiation has a much greater penetrating power in matter than beta particles and consequently presents a serious shielding problem in any situation where it is produced in large quantities. A nuclear reactor is an example of this and further consideration of the characteristics, range and attenuation of gamma radiation in matter will be left to the chapter devoted to radiation hazards and reactor shielding.

Chapter 2

Neutrons and their interaction with matter

The operation of a nuclear reactor depends entirely on the way in which neutrons interact with the atomic nuclei of the materials in the reactor. It is therefore necessary to study these interactions in sufficient detail to develop an understanding of the processes which take place in a reactor.

In this chapter we shall describe the characteristics of neutrons and those neutron interactions which are of interest to the nuclear engineer, with particular emphasis on the fission process which is of central importance. The ideas of neutron flux and neutron cross-sections will be introduced in order to express interaction rates quantitatively, and the variation of cross-sections with neutron energy will be described. Finally, we shall consider the scattering of high energy neutrons resulting in their progressive loss of energy, and the situation that exists when these neutrons lose sufficient energy to reach thermal equilibrium with their surroundings.

The subject matter of this chapter can be regarded as the foundation upon which the description of nuclear reactors and the theory of their operation, which is dealt with in the following chapters, is based.

2.1 Neutrons and neutron sources

The discovery of the neutron took place during the years between 1920 and 1932, and culminated in 1930 in the discovery of a penetrating radiation produced by the interaction of alpha particles and light elements such as beryllium. This was explained in 1932 by Chadwick as being particles of mass nearly equal to that of the proton and zero charge. The name neutron was given to these particles.

Neutrons do not exist in nature, and those neutrons which are produced by one of the processes to be described presently decay by transforming into a proton and an electron. The average lifetime of a neutron is about 12 minutes; however, as the average lifetime of a neutron in a reactor is invariably only a small fraction of a second, the natural decay of neutrons does not concern us.

Nuclear reactions are the only sources of neutrons, and if we discount fission as a source of neutrons for the time being the most common sources depend on (α, n) and (γ, n) reactions with light elements. For example:

$$^9\text{Be} + {}^4\text{He} \rightarrow {}^{12}\text{C} + {}^1n$$

and

$$^2\text{H} + \gamma \rightarrow {}^1\text{H} + {}^1n$$

A very common neutron source makes use of the $^9\text{Be}(\alpha, n) {}^{12}\text{C}$ reaction. The alpha particles which are required to cause the reaction are obtained from a naturally occurring alpha-emitting radioactive isotope such as radium 226. Thus a mixture of one gramme of radium and several grammes of beryllium provides a compact source emitting about 10^7 neutrons per second.

2.2 Neutron interactions with atomic nuclei

One obvious and important feature of neutron–nucleus interactions is that the neutron, being uncharged, does not undergo a force of repulsion as it approaches a nucleus. It is therefore possible for neutrons of any energy to interact with nuclei, and as Fermi discovered, low energy neutrons usually interact more readily than high energy neutrons.

There are two principal mechanisms of interaction between neutrons and nuclei, namely potential scattering and compound nucleus formation. The latter process may result in the neutron being captured in the nucleus, being effectively scattered, or causing the fission or splitting of the nucleus.

2.3 Potential scattering

Potential scattering is a process in which the incident neutron is scattered or bounced off the nucleus, and it can be regarded as similar to the collision of two billiard balls of unequal mass. The nucleus remains unchanged and at its ground state of energy during the process which involves only a transfer of kinetic energy between the neutron and the nucleus. Often the kinetic energy of the incident neutron is much greater than that of the nucleus and as a result of the interaction the neutron emerges with reduced kinetic energy. The laws of conservation of kinetic energy and conservation of momentum are both valid for this type of interaction, which for this reason is often called elastic scattering.

Consider a neutron of mass m moving with velocity v_1 towards a nucleus of mass M. In many interactions the speed of the neutron is much greater than that of the nucleus, and the latter can be assumed to be at rest. The situation as viewed in the laboratory (L system) is shown in Figure 2.1(a). We will now adopt a system of coordinates in which the centre-of-mass of the neutron–nucleus pair is at rest. In the L system the centre-of-mass is moving towards the nucleus with a velocity given by:

$$v_c = \frac{m}{M + m} v_1 \qquad (2.1)$$

To bring the centre-of-mass to rest it is necessary to impose a velocity equal in magnitude and opposite in direction to v_c on the whole system which becomes transformed to that shown in Figure 2.1(b). The neutron and nucleus are now depicted in the centre-of-mass system (C system) of coordinates and their velocities are respectively:

$$u_1 = \frac{M}{M + m} v_1 \qquad (2.2)$$

and

$$U_1 = - \frac{m}{M + m} v_1 \qquad (2.3)$$

Notice that:

$$mu_1 + MU_1 = 0 \qquad (2.4)$$

This equation expresses the fact that the momentum of the neutron–nucleus pair is zero in the C system, and it will remain zero after any reaction for which the law of conservation of momentum is valid.

The total kinetic energy, E_c, of the two particles in the C system prior to any interaction is given by:

$$
\begin{aligned}
E_c &= \tfrac{1}{2}mu_1^2 + \tfrac{1}{2}MU_1^2 \\
&= \frac{1}{2}\frac{mM^2}{(m + M)^2}v_1^2 + \frac{1}{2}\frac{Mm^2}{(m + M)^2}v_1^2 \\
&= \frac{1}{2}\frac{mM}{(m + M)}v_1^2 \\
&= \frac{M}{m + M}E_1 \qquad (2.5)
\end{aligned}
$$

where E_1 is the kinetic energy of the neutron in the L system. It is evident that the kinetic energy of the two particles in the C system is always somewhat less than in the L system.

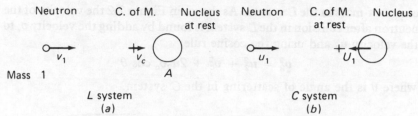

Figure 2.1. Neutron collisions in the L and C systems

Bearing in mind that atomic masses are very nearly the same as mass numbers, equations (2.1) to (2.5) can be written alternatively as:

$$v_c = \frac{1}{A+1} v_1 \tag{2.1}$$

$$u_1 = \frac{A}{A+1} v_1 \tag{2.2}$$

$$U_1 = -\frac{1}{A+1} v_1 \tag{2.3}$$

and

$$E_c = \frac{A}{A+1} E_1 \tag{2.5}$$

After an elastic scattering collision the momentum of the neutron–nucleus pair in the C system is zero and their kinetic energy is unchanged. In order that the momentum remains zero the neutron and nucleus must move away from the site of the collision in opposite directions with velocities u_2 and U_2 respectively such that:

$$mu_2 + MU_2 = 0 \tag{2.6}$$

To conserve kinetic energy:

$$\tfrac{1}{2}mu_1^2 + \tfrac{1}{2}MU_1^2 = \tfrac{1}{2}mu_2^2 + \tfrac{1}{2}MU_2^2 \tag{2.7}$$

Using equations (2.4) and (2.6) to eliminate u_1 and u_2 from (2.7) we get:

$$U_1 = U_2 \tag{2.8}$$

From this it follows that:

$$u_1 = u_2 \tag{2.9}$$

These equations express the simple and important fact that in an elastic scattering collision the speeds of neutron and nucleus in the C system are unchanged, regardless of the angle of scattering.

To find the effect of the collision on the neutron speed in the L system (the real state of affairs as far as we are concerned), we must revert to that system by imposing on the C system a velocity v_c, the velocity of the centre-of-mass in the L system. As shown in Figure 2.2 the velocity of the neutron after collision in the L system is found by adding the velocity v_c to the velocity u_2 and using the cosine rule:

$$v_2^2 = u_2^2 + v_c^2 + 2u_2v_c \cos.\theta$$

where θ is the angle of scattering in the C system.

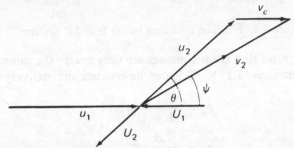

Figure 2.2. Neutron velocity after elastic scattering

Using equations (2.1), (2.2) and (2.9) to obtain u_2 and v_c in terms of v_1, we have:

$$v_2^2 = v_1^2 \left\{ \frac{A^2 + 1 + 2A \cos \theta}{(A + 1)^2} \right\}$$

or

$$E_2 = E_1 \left\{ \frac{A^2 + 1 + 2A \cos \theta}{(A + 1)^2} \right\} \qquad (2.10)$$

Equation (2.10) shows that the neutron speed or energy after elastic scattering is a function of the energy before collision, the mass number of the scattering nucleus and the angle of scattering in the C system. Let us consider two extreme cases. Firstly, if a glancing collision occurs in which the neutron is barely deflected, $\theta = 0$, and $E_2 = E_1$. The neutron energy is unaltered as a result of a glancing collision.

Secondly, if $\theta = 180°$, corresponding to a head-on collision as a result of which the neutron bounces back along its original path, $\cos \theta = -1$. For this case equation (2.10) becomes:

$$E_2 = E_1 \times \left(\frac{A - 1}{A + 1} \right)^2 \qquad (2.11)$$

or $E_2 = \alpha E_1$, where

$$\alpha = \left(\frac{A - 1}{A + 1} \right)^2 \qquad (2.12)$$

The fractional loss of energy is given by

$$\frac{E_1 - E_2}{E_1} = (1 - \alpha)$$

This represents the maximum fractional loss of energy that a neutron can suffer as a result of an elastic scattering collision; it is a function of the mass number of the scattering nucleus and as the mass number decreases, the maximum fractional loss of energy increases. In the limiting case for hydrogen, $A = 1$, $\alpha = 0$ and a neutron can lose all its energy in a single head-on collision.

2.4 Compound nucleus formation

The process of compound nucleus formation involves firstly the absorption of the incident neutron into the original nucleus to form a compound nucleus:

$$^A_Z X + {}^1_0 n \rightarrow {}^{A+1}_Z X$$

When the compound nucleus is formed it must be at rest in the C system to conserve zero momentum, and so its kinetic energy is also zero. It follows that the kinetic energy of the neutron–nucleus pair before the reaction in the C system is transformed into excitation energy of the compound nucleus. This is the energy E_c given by equation (2.5). The total excitation energy of the compound nucleus is $E_c + B$, where B is the binding energy of a neutron in the nucleus of ^{A+1}X. For example when a low energy neutron is absorbed in boron 10 to form boron 11 the latter is formed at 11·4 MeV above its ground state, as can be verified by consulting the data of Tables 1.1 and 1.2 in Chapter 1. If the incident neutron had an energy of 1 MeV the boron 11 would be formed at 12·3 MeV above its ground state.

2.5 Capture

As pointed out in Chapter 1, the excited compound nucleus decays immediately and emits either a particle or gamma radiation or both, and it is the product of the decay that distinguishes one type of compound nucleus interaction from another.

If the compound nucleus decays to its ground state by the emission of gamma radiation only, then the complete interaction is called a capture, or sometimes radiative capture or (n, γ) reaction because the effect is to capture the original neutron and emit gamma radiation. Two examples of this reaction are:

$$^{23}\text{Na} + {}^1 n \rightarrow {}^{24}\text{Na}^* \rightarrow {}^{24}\text{Na} + \gamma, \quad \text{or} \quad {}^{23}\text{Na}(n, \gamma) {}^{24}\text{Na}$$

and $\quad ^{238}\text{U} + {}^1 n \rightarrow {}^{239}\text{U}^* \rightarrow {}^{239}\text{U} + \gamma, \quad \text{or} \quad {}^{238}\text{U}(n, \gamma) {}^{239}\text{U}$

(An asterisk denotes the compound nucleus in an excited state.)

If the excited compound nucleus decays by the emission of an alpha particle, the reaction is known as an (n, α) reaction. If the product nucleus after the emission of the alpha particle is above its ground state of energy, it will decay to the ground state by the emission of gamma radiation. An example of such a reaction is the boron 10–neutron interaction described in Chapter 1:

$$^{10}B + {}^1n \rightarrow {}^{11}B^* \rightarrow {}^7Li + {}^4He, \quad \text{or} \quad {}^{10}B(n, \alpha) {}^7Li$$

2.6 Inelastic scattering

If the compound nucleus decays by the emission of a neutron, the interaction is effectively a scattering process, although the emitted neutron is not necessarily the same as the incident neutron. If the nucleus resulting from the emission of the neutron (which is of course identical with the original nucleus) is still at an excited state of energy, it decays to its ground state by the emission of gamma radiation, and a process of this type is referred to as inelastic scattering. The law of conservation of kinetic energy is not valid for inelastic scattering since some of the original kinetic energy is transformed to gamma radiation, and the total kinetic energy is less than it was before the reaction. An important characteristic of this reaction, which only takes place between neutrons of fairly high energy and intermediate and heavy mass nuclei, is that neutrons lose on average much more energy per collision than in the case of elastic scattering with the same nucleus.

If after the emission of a neutron from the compound nucleus the original nucleus is formed at its ground state of energy, then kinetic energy is conserved in the reaction and it is called compound elastic scattering. Clearly, as far as the results of the interactions are concerned, potential scattering and compound elastic scattering can be regarded as equivalent, and the sum of these two processes is usually referred to as elastic scattering.

In the case of a few of the heaviest elements the compound nucleus may be formed in such a state of excitation that it decays by splitting into two intermediate nuclei of unequal mass. This is nuclear fission and it will be considered in detail in the next section.

2.7 Fission

The discovery of fission was made in Germany in 1938 by Hahn and Strassmann who were studying the radioactive isotopes formed as a result of the bombardment of uranium by neutrons in an effort to produce transuranium elements. One of the elements identified in the products of the reactions was radioactive barium 139, which indicated a hitherto unknown type of reaction in which the uranium nucleus split into fragments which were themselves nuclei of intermediate mass elements. Further work showed the presence of several other elements of medium mass number, and the existence of the fission process was definitely established. Shortly afterwards it was shown that neutrons were also emitted in the process and the possibility of a chain reaction was realized in which neutrons emitted in one fission event might be able to cause further fission, thus establishing a continuous reaction.

The isotope of uranium that is principally responsible for fission is uranium 235, which is present in naturally occurring uranium to the extent of 0·715 per cent. In this isotope fission can be caused by neutrons of any energy, low energy neutrons being the most effective. Fission in uranium 238, which comprises 99·285 per cent of natural uranium, can only be caused by neutrons of energy greater than 1 MeV.

There are three other isotopes of importance which can undergo fission. Thorium 232, the only naturally occurring isotope of that element, is fissionable with neutrons with energy greater than about 1·4 MeV, and two isotopes, uranium 233 and plutonium 239, which do not occur naturally but can be produced artificially by nuclear reactions, undergo fission with neutrons of all energies, low energy neutrons being again the most effective. It is customary to refer to the five isotopes mentioned above (and any other isotopes which undergo fission with neutrons of energy less than about 10 MeV) as fissionable, and to reserve the term fissile for the three isotopes ^{233}U, ^{235}U and ^{239}Pu which undergo fission with low energy neutrons.

The theory of fission is beyond the scope of this book, however a brief description of the generally accepted liquid drop model will give a qualitative picture of the processes involved. The short range nuclear forces, which are analogous to the surface tension of a liquid drop, hold the nucleus in a more or less spherical shape in the same way that a liquid drop is spherical, however if the nucleus is excited, possibly as a result of absorbing a neutron, its shape may be distorted. In most cases the distortion is limited by the action of the nuclear forces and after de-excitation the spherical shape of the nucleus is restored, however it is possible that the distortion may lead to a dumbbell shape at which point the Coulomb force of repulsion between the two halves of the dumbbell exceeds the nuclear force which is weakened by the

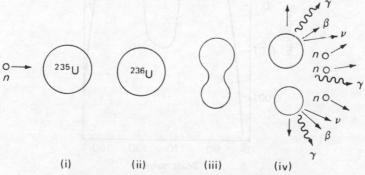

$$ \text{(i)} \qquad \text{(ii)} \qquad \text{(iii)} \qquad \text{(iv)} $$

Figure 2.3. Four stages in the fission process.

distortion of the nucleus. Once this point is reached the nucleus splits into two fragments.

The characteristics of fission will be described by considering uranium 235, however the fission of the other four isotopes is essentially the same in all respects. The first stage of the reaction is the absorption of a neutron in ^{235}U to form ^{236}U at an excited state. In some cases the ^{236}U goes to its ground state of energy by the emission of gamma radiation, an example of an (n, γ) reaction, however in the majority of cases the ^{236}U nucleus splits as described above. The products of fission are two fission fragments whose mass numbers vary between about 70 and 160, a number of neutrons varying between none and five, beta particles, gamma radiation, neutrinos and energy. These products are shown diagrammatically in Figure 2.3.

The exact identity of the fission products and the number of neutrons vary from one fission event to another, however the following reaction is typical:

$$^{235}_{92}U + {}^{1}_{0}n \rightarrow {}^{236}_{92}U^* \rightarrow {}^{147}_{57}La + {}^{87}_{35}Br + 2{}^{1}_{0}n$$

It will be seen that the masses of the two fission products, lanthanum and bromine in this example, are not the same, and unsymmetrical fission such as this is much more likely than fission with two products of equal mass. The fission product spectrum for ^{235}U is shown in Figure 2.4, and it can be clearly seen that the mass numbers of all the fission products lie between 70 and 160, that the most probable mass numbers which occur in about 6·5 per cent of fissions are about 96 and

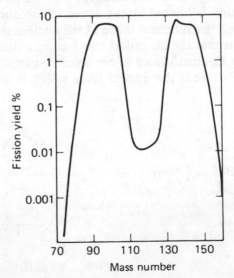

Figure 2.4. The fission product spectrum for uranium 235

135, and that symmetrical fission with two products of mass 117 only occurs about once in every 20 000 fissions.

The fission products are all radioactive, which is to be expected since if a very heavy nucleus (in which the neutron to proton ratio is just over 3 to 2) splits into two intermediate mass nuclei (in which the neutron to proton ratio for stability is just under 3 to 2), the fission products have a surplus of neutrons. As we have seen, such nuclei decay mainly by beta particle emission, although in a very few cases the decay is by neutron emission. In some cases long decay chains are formed, an example for the fission product tellurium 135 being:

$$^{135}\text{Te} \xrightarrow{\beta} {}^{135}\text{I} \xrightarrow{\beta} {}^{135}\text{Xe} \xrightarrow{\beta} {}^{135}\text{Cs} \xrightarrow{\beta} {}^{135}\text{Ba}$$

The radioactivity of the fission products creates serious hazards and shielding problems in reactors, particularly in the handling of irradiated uranium fuel. Another problem that arises in the case of a very few fission products is the build-up in the reactor of isotopes which capture neutrons to a large extent and which even in small quantities have a serious effect on the possibility of establishing a continuous fission reaction. Xenon 135, a daughter product in the decay chain illustrated above, is the most notable example of this, and its effect on the design and operation of a nuclear reactor will be considered in detail in a later chapter.

The majority of neutrons emitted in the fission process are released at the instant of fission, and are known as prompt neutrons. As mentioned above, a few of the fission products decay by neutron emission and this provides a further source of neutrons which are released sometime after the original fission event, depending on the half-life of the radioactive fission product concerned. An example of a neutron-emitting decay chain is shown in Figure 2.5.

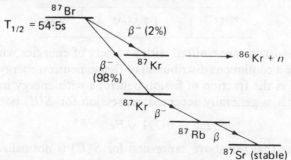

Figure 2.5. The decay of the fission product bromine 87

The fission product bromine 87 decays by beta particle emission with a half-life of 54·5 seconds to form krypton 87. In 2 per cent of cases the ^{87}Kr is formed at such an excited state that it decays immediately by neutron emission to form ^{86}Kr. This is therefore a source, albeit a very small one, of neutrons whose emission is delayed by an average of about 80 seconds (the mean life of ^{87}Br) after the fission event that was their origin. The total fraction of neutrons which are delayed is only (in the case of ^{235}U) about 0·65 per cent of the total neutron yield, however they play a vital role in the control of nuclear reactors as we shall see later.

The number of neutrons emitted per fission varies from one event to another, and also depends on the isotope undergoing fission and the energy of the incident neutrons. The average number of neutrons emitted per fission, ν, is one of the most important parameters in nuclear engineering. The variation of ν with energy is more or less

$$\nu(E) = \nu_0 + \alpha E$$

and some values are given in Table 2.1.

Table 2.1. The number of neutrons emitted per fission

Isotope	Incident neutron energy	ν
^{235}U	0·025 eV	2·42
	1 MeV	2·51
^{239}Pu	0·025 eV	2·93
	1 MeV	3·04
^{233}U	0·025 eV	2·49
	1 MeV	2·58
^{232}Th	1·5 MeV	2·12
^{238}U	1·1 MeV	2·46

Fission neutrons are emitted with a variety of energies which can be described by a continuous distribution or fission neutron energy spectrum. If $S(E)\, dE$ is the fraction of fission neutrons with energy in the range E to $E + dE$, a generally accepted expression for $S(E)$ is:

$$S(E) = 0·771 \sqrt{E} e^{-0·776E} \tag{2.13}$$

(Note that the above expression for $S(E)$ is normalized, so that

$$\int_0^\infty S(E) dE = 1)$$

The average energy of fission neutrons is found from the equation:

$$\bar{E} = \frac{\int_0^\infty ES(E)\, dE}{\int_0^\infty S(E)\, dE}$$

and is 1·93 MeV, however it is usually taken as 2 MeV. Figure 2.6 shows the fission neutron energy spectrum.

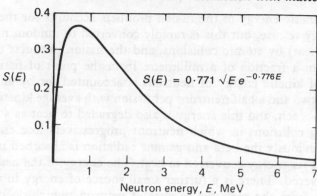

Figure 2.6. The fission neutron energy spectrum

Gamma radiation is emitted both promptly at the instant of fission and at a later stage during the decay of fission products. Beta radiation and neutrinos are emitted during fission product decay. The gamma radiation, with its high penetrating power, presents a shielding problem, however reactors are usually large enough that most of the gamma radiation is absorbed within them. Beta radiation is also absorbed within the reactor, but neutrinos stream out and their energy is lost, however they are not a hazard to health or safety.

The energy release as a result of fission has already been mentioned in connection with the binding energy curve. A very approximate calculation based on this curve gives some idea of the magnitude of this energy. The binding energy per nucleon of ^{235}U is about 7·6 MeV, and the binding energy per nucleon of nuclei of mass number in the region of 117 is about 8·5 MeV, so the energy release is approximately $235 \times (8\cdot5–7\cdot6)$ which is about 212 MeV. Much more accurate measurements of the energy release have been made and are summarized in Table 2.2.

Table 2.2. *The emitted and recoverable energy for the fission of* ^{235}U

	Emitted energy (MeV)	Recoverable energy (MeV)
Fission products	168	168
Fission neutrons	5	5
Prompt γ radiation	7	7
Fission product decay		
$\quad\beta$ radiation	8	8
$\quad\gamma$ radiation	7	7
$\quad$neutrinos	12	0
Capture γ radiation	0	~5
Total	207	~200

The kinetic energy of the fission products accounts for the bulk of the energy release, but this is rapidly converted to random molecular energy (heat) by atomic collisions, and the fission products travel no more than a fraction of a millimetre from the point of fission. The 5 MeV of kinetic energy of neutrons is accounted for by an average of about two and a half neutrons per fission with average kinetic energy of 2 MeV each, and this energy is also degraded to heat as a result of scattering collisions in which neutrons progressively lose energy. As noted previously the beta and gamma radiation is absorbed in the reactor and its energy converted to heat. The energy of the neutrinos is not recovered. There is a further small source of energy in a reactor resulting from the radiative capture of neutrons which do not cause fission, the capture gamma radiation which is emitted is for the most part absorbed and its energy converted to heat. The figure of 200 MeV of recoverable energy per fission is not exact, however it is the generally accepted value.

2.8 Neutron flux, cross-sections and interaction rates

It is now necessary to establish a framework for measuring neutron interaction rates quantitatively. These rates do not depend on the direction of neutron motion within the target material so we may visualize for simplicity a situation, which seldom occurs in practice, in which all neutrons are moving in the same direction in a parallel beam. The neutron flux, ϕ, may be defined as the total number of neutrons which pass through a unit area normal to their direction per second.

If all the neutrons have the same speed v, and if the neutron density is n neutrons per unit volume, then:

$$\phi = nv$$

If in the more likely event the neutrons have a spectrum of speeds such that $n(v)\,dv$ is the number of neutrons per unit volume whose speed is in the range v to $v + dv$, then:

$$\phi = \int\limits_{\substack{\text{all}\\\text{speeds}}} vn(v)\,dv$$

For the case in which neutrons are moving in all directions the neutron flux can be defined as the total track length of all neutrons in a unit volume per second. This definition is consistent with the one given a few lines above for a parallel beam of neutrons, but it does not depend on that condition. Being applicable to neutrons moving randomly in all directions without reference to their direction of motion, it emphasizes the scalar (as opposed to vector) nature of the neutron flux.

The interaction rate between a beam of neutrons and the nuclei in a target material has been experimentally observed to be proportional (*a*) to the neutron flux, and (*b*) to the number of atoms in the target, which is assumed to consist of a single isotope.

Consider a beam of neutrons, all of speed v cm/s and density n neutrons/cm^3, incident on a target of area A cm^2 and thickness dx cm containing N nuclei/cm^3, see Figure 2.7.

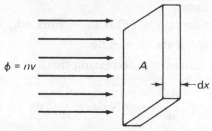

Figure 2.7. The interaction rate of neutrons

The preceding statement can now be used to express the interaction rate F in the target material thus:

$$F \propto \phi NV$$
$$F = \sigma \phi NV \tag{2.14}$$

where $V = A$ dx, the volume of the target, and NV is the total number of atoms of the isotope in the target in which the reaction is taking place.

The constant σ in equation (2.14) is known as the microscopic cross-section of the isotope concerned. Its units are cm^2/nucleus, and it can be regarded as the area presented by each nucleus to neutrons to cause a reaction. (This area is not the same as the actual size of the nucleus, in some cases it may be larger, in others it may be smaller.) The values of σ for most isotopes lie between 10^{-22} and 10^{-26} cm^2, and the usual unit in which values of σ are quoted is the barn.

$$1 \text{ barn} = 10^{-24} \text{ cm}^2$$

The total cross-section of all the nuclei in unit volume of a material is called the macroscopic cross-section, Σ, and has units cm^2/cm^3 or cm^{-1},

$$\Sigma = N\sigma$$

and the interaction rate per unit volume,

$$F = \Sigma \phi \tag{2.15}$$

The probability that a neutron entering the target will interact within a distance dx is:

$$\frac{\text{Number of neutrons interacting per second in a target of thickness d}x}{\text{Number of neutrons per second incident on the target}} = \frac{\sigma \phi NA \text{ d}x}{\phi A}$$

$$= \sigma N \text{ d}x$$

$$= \Sigma \text{d}x \tag{2.16}$$

It follows that the macroscopic cross-section can be interpreted as the probability per unit track length that a neutron will interact.

The attenuation of a beam of neutrons in a target material can be found by equating the rate of interaction in an element of thickness dx to the difference between the number of neutrons entering and leaving the element per second:

$$\text{Interaction rate} = (\text{Flux in} - \text{Flux out}) \times A$$

$$\sigma \phi N A\ dx = -A\ d\phi$$

(the negative sign indicates a decreasing flux).

Rearranging:

$$\frac{d\phi}{\phi} = -\sigma N\ dx = -\Sigma\ dx$$

The solution of this equation for $\phi(x)$, the flux which penetrates to a distance x without interacting, is:

$$\phi(x) = \phi_0\ e^{-\Sigma x} \tag{2.17}$$

where ϕ_0 is the incident neutron flux.

The average distance that a neutron travels without interacting is known as the mean free path, λ. Considering a number of neutrons, n, we may write:

$$\lambda = \frac{1}{n} \sum_{\substack{\text{all} \\ \text{neutrons}}} \left(\begin{array}{c} \text{Number of neutrons} \\ \text{which travel a distance} \\ x \text{ without interacting} \end{array} \right) \times \left(\begin{array}{c} \text{Probability of} \\ \text{interaction in} \\ \text{distance } dx \end{array} \right) \times x$$

The quantity on the right of the summation is the total distance travelled by all neutrons without interaction. Using equations (2.16) and (2.17) and altering the right-hand side to an integral over all possible values of x, namely 0 to ∞, the equation for λ is:

$$\lambda = \frac{1}{n} \int_0^\infty n\ e^{-\Sigma x}\ \Sigma x\ dx$$

$$= \frac{1}{\Sigma} \tag{2.18}$$

Thus the mean free path is seen to be the reciprocal of the macroscopic cross-section.

The preceding discussion in this section has made no distinction between the different types of neutron interaction such as scattering, capture or fission, and all the statements and results obtained so far apply to any type of interaction.

Thus the rates at which elastic scattering, inelastic scattering, capture and fission take place are characterized by the elastic scattering cross-section, σ_s, the inelastic scattering cross-section, σ_i, the capture cross-section, σ_c, and the fission cross-section, σ_f, (which is zero for all non-fissionable isotopes). The total cross-section, σ_t, is the sum of these cross-sections, and measures the rate at which any type of interaction takes place:

$$\sigma_t = \sigma_s + \sigma_i + \sigma_c + \sigma_f$$

The absorption cross-section, σ_a, is the sum of the capture and fission cross-sections, and is of course, the same as the capture cross-section for all non-fissionable isotopes:

$$\sigma_a = \sigma_c + \sigma_f$$

In addition to the use of cross-sections as a means of determining reaction rates, a qualitative interpretation as a measure of probability is often useful. For example, if an isotope has capture and elastic scattering cross-sections of 0·1 and 10 barns respectively, it is evident that elastic scattering is the most probable reaction in this material and that by comparison capture is almost negligible, occurring in less than 1 per cent of all reactions.

The distinction between mean free paths may be made by defining the scattering mean free path, λ_s, as the average distance travelled by neutrons between scattering collisions:

$$\lambda_s = \frac{1}{\Sigma_s}$$

The absorption mean free path, λ_a, is the average distance travelled by neutrons up to the point where they are absorbed:

$$\lambda_a = \frac{1}{\Sigma_a}$$

The total mean free path, λ_t, is given by the equation:

$$\frac{1}{\lambda_t} = \frac{1}{\lambda_a} + \frac{1}{\lambda_s}$$

It is frequently necessary in nuclear engineering to calculate cross-sections for compounds and mixtures of materials. Strictly speaking a microscopic cross-section can only refer to a single isotope, however it is often convenient to use an average microscopic cross-section for a naturally occurring element which consists of a mixture of isotopes.
Example: Calculate the microscopic absorption cross-section of natural uranium, which consists of 99·285 per cent ^{238}U and 0·715 per cent ^{235}U. The microscopic cross-sections for 0·025 eV neutrons are:

^{238}U: $\sigma_c = 2.72$ barns ^{235}U: $\sigma_c = 101$ barns

$\quad\quad\quad \sigma_f = 0$ $\quad\quad\quad\quad \sigma_f = 579$ barns

σ_a for natural uranium $= 0.99285 \times 2.72 + 0.00715 \times 680$

$\quad\quad\quad\quad\quad\quad\quad\quad\quad\quad = 7.6$ barns

Macroscopic cross-sections for mixtures and compounds can be calculated from a knowledge of the cross-sections and numbers of atoms per unit volume of the constituents. If a mixture or compound contains N_1, N_2, N_3, etc. atoms per cm^3 of elements whose microscopic cross-sections are $\sigma_1, \sigma_2, \sigma_3$, etc., the macroscopic cross-section of the mixture or compound, Σ, is given by:

$$\Sigma = N_1\sigma_1 + N_2\sigma_2 + N_3\sigma_3 + \cdots$$

This equation is based on the assumption that the nuclei of the elements in the compound interact with neutrons independently of each other. If the molecular or crystalline structure affects the neutron interaction processes the equation is not valid. Two important examples of this effect are the elastic scattering of low energy neutrons by water and heavy water which will be referred to in the next section.

In the case of a compound, the numbers of atoms per cm^3 of each of the constituents can be determined from a knowledge of the density of the compound, its molecular weight and Avogadro's Number.

Example: Calculate the macroscopic capture cross-section of water of density 1 g/cm^3. The microscopic capture cross-sections of hydrogen and oxygen are 0·332 barns and 0·0002 barns respectively.

Using Avogadro's number, the number of molecules of water per cm^3 is:

$$\frac{0\cdot001 \times 6\cdot023 \times 10^{26}}{18} = 3\cdot35 \times 10^{22}$$

$$\begin{aligned}\Sigma_c \text{ for water} = {}& 2 \times 3\cdot35 \times 10^{22} \times 0\cdot332 \times 10^{-24} \\ & + 3\cdot35 \times 10^{22} \times 0\cdot0002 \times 10^{-24} \\ = {}& 0\cdot0222 \text{ cm}^{-1}\end{aligned}$$

2.9 The variation of cross-sections with neutron energy

Cross-sections for neutron interactions are in many cases not constant, but vary with neutron energy. A complete description of the variation of cross-sections is beyond the scope of this book, and we will restrict ourselves to a general view, with a more detailed look at some examples of particular interest in nuclear engineering. In general the variation of cross-section depends on the type of interaction involved, whether scattering or absorption, and the mass number of the element involved. For reasons that will be clear later we will restrict ourselves to neutron energies between 0·01 eV and 10 MeV.

Elastic scattering cross-sections for light elements are more or less independent of neutron energy up to about 1 MeV. For intermediate and heavy elements the elastic scattering cross-section is constant at low energy and exhibits some variation at higher energy. However, we are usually more interested in light elements as far as elastic scattering is concerned so as a generalization we may regard σ_s as being constant at all energies for all elements of interest. Furthermore, there is not a great deal of variation from one element to another, and nearly all elements have scattering cross-sections in the range 2 to 20 barns. The important exceptions to this concern water and heavy water in which the molecular structure affects the scattering of low energy neutrons in such a way that although σ_s for the free atoms of hydrogen, deuterium and oxygen remain constant, the value of σ_s for water and heavy water rises as the neutron energy falls below 1 eV. Figure 2.8 illustrates this effect.

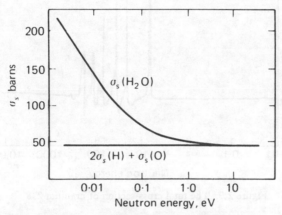

Figure 2.8. The elastic scattering cross-section of water for low energy neutrons

Inelastic scattering occurs principally between high energy neutrons and intermediate and heavy elements, and is of importance in nuclear engineering because high energy neutrons can lose a large fraction of their energy as a result of inelastic scattering with heavy elements such as uranium. Inelastic scattering with light nuclei is not of much importance because the threshold energy below which $\sigma_i = 0$ is very high. The threshold energies for inelastic scattering for oxygen, sodium and uranium are about 6·5 MeV, 0·4 MeV and 0·05 MeV respectively, and above these thresholds the inelastic scattering cross-section rises to a more or less constant and rather small value, generally a few barns.

Absorption cross-sections exhibit much more variation than elastic scattering cross-sections, not only from one isotope to another but also with varying neutron energy. The cross-section for many light isotopes is inversely proportional to the neutron speed over a wide range of energies, i.e.

$$\sigma_a \propto \frac{1}{v} \propto \frac{1}{\sqrt{E}}$$

The variation for boron 10 is shown in Figure 2.10.

For heavy isotopes the $1/v$ variation is exhibited at low energies up to about 10 eV. In the intermediate energy range from about 10 eV to 1000 eV the cross-section displays a very erratic behaviour and rises to a number of peaks known as resonances at which the cross-section values may be very large. At high energies above 1000 eV the resonances cannot be resolved and the cross-section assumes a fairly constant value of a few barns. The variation of σ_t for ^{238}U is shown in Figure 2.9.

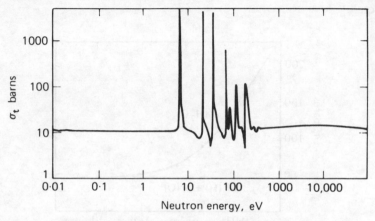

Figure 2.9. The total cross-section of uranium 238

Isotopes of intermediate mass exhibit a behaviour between that of the light and heavy isotopes, thus a typical variation would be $\sigma_a \propto 1/v$ at low energies, then a few resonances at intermediate energies, and

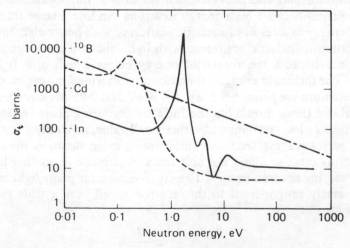

Figure 2.10. The total cross-sections of cadmium, indium and boron 10

finally a rather low and slightly varying value at high energies. The cross-sections of cadmium and indium are shown in Figure 2.10.

It is possible to explain qualitatively some of the reasons for the characteristic variations of absorption cross-sections. At energies below the resonance energy region the probability of interaction is governed by the time during which the neutron is in the neighbourhood of the nucleus, and this time varies inversely as the neutron speed. Thus we have the variation expressed by $\sigma_a \propto 1/v$.

To explain the resonance peaks it should be borne in mind that we are dealing with interactions involving compound nucleus formation. The compound nucleus is formed with an excitation energy of $B + E_c$, where B is the binding energy of a neutron in the compound nucleus and E_c is the neutron energy multiplied by $A/(A + 1)$. If the energy of the neutron is such as to produce the compound nucleus at or very near one of its excited states as shown in Figure 2.11 the probability of the interaction taking place is very high, corresponding to a high value of the cross-section. If, on the other hand, the energy of the incident neutron is such as to produce the compound nucleus at some energy midway between two excited levels, the probability of the interaction is very low, corresponding to a low value of the cross-section. Heavy nuclei such as ^{239}U have a large number of closely spaced excited levels which accounts for the large number of resonances in the absorption cross-section of ^{238}U.

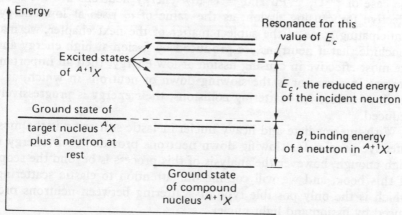

Figure 2.11. Energy levels and resonance absorption

The fission cross-sections of ^{235}U, ^{233}U and ^{239}Pu vary in much the same way as capture cross-sections of heavy isotopes described above. Figure 2.12 shows the variation of σ_f for ^{235}U which exhibits the characteristic $1/v$ portion at low energy, resonances at intermediate energy, and the smooth curve a high energy where individual resonances overlap. The fission cross-sections of ^{238}U and ^{232}Th show the

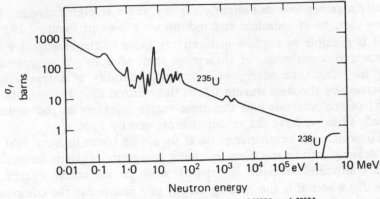

Figure 2.12. Fission cross-sections of ^{235}U and ^{238}U

existence of a threshold neutron energy below which fission does not take place. Above the threshold σ_f rises to a more or less constant and rather small value, see Figure 2.12 for the fission cross-section of ^{238}U.

2.10 Elastic scattering and slowing down of neutrons

In preceding sections of this chapter we have seen that neutrons are produced by fission at an average energy of about 2 MeV, and that in the case of ^{235}U, ^{239}Pu and ^{233}U low energy neutrons are the most effective for causing fission, as the value of σ_f rises at low energy. Anticipating some of the subject matter of the next chapter, we may conclude that if neutrons are produced by fission at high energy and are most effective in causing fission at low energy, then an important process in a reactor is the slowing down of neutrons in which, as a result of successive scattering collisions, their energy is progressively reduced.

With intermediate and heavy nuclei inelastic scattering is the most effective process for slowing down neutrons provided their energy is high enough, however the analysis of this process is beyond the scope of this book, and we will confine our attention to elastic scattering, which is the only possible type of scattering between neutrons produced by fission and light nuclei.

In an earlier section of this chapter, the analysis of elastic scattering of neutrons was carried to the point where it was shown that, according to equation (2.10), the energy of a neutron after an elastic scattering collision is related to its energy before collision, the mass of the scattering nucleus and the angle of scattering in the C System. In a reactor in which billions of neutrons are being scattered each second, we are interested not so much in the loss of energy of each neutron as the average loss of energy per scattering collision of all these neutrons, which depends on their angular distribution after scattering. In the C system this distribution is given by the empirical scattering law:

Elastic scattering is isotropic in the C system.

In other words there is no preferential direction of scattering in the C system. If we imagine a large number of neutrons, all moving initially along the same path and being scattered at point O, the centre of a sphere, then the number of neutrons emerging through any unit area on the surface of the sphere is independent of the position of the unit area. We can deduce a scattering law in terms of θ,, the angle of scattering in the C system.

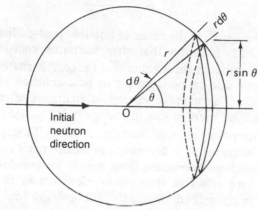

Figure 2.13. Scattering law for the C system

Referring to Figure 2.13, the fraction of neutrons whose angle of scattering in the C system is between θ and $\theta + d\theta$ is given by the distribution function $p(\theta)\, d\theta$:

$$p(\theta)\, d\theta = \frac{\text{The area of the ring subtended on the surface of the sphere by the angle } d\theta}{\text{Total surface area of the sphere}}$$

$$= \frac{2\pi r \sin \theta\, r\, d\theta}{4\pi r^2} = \tfrac{1}{2} \sin \theta\, d\theta \qquad (2.19)$$

It is more useful to have a distribution function in terms of the post-collision energy. If $P(E)\, dE$ is the fraction of post-collision neutrons with energies in the range E to $E + dE$, then $p(\theta)\, d\theta$ and $P(E)\, dE$ are related by:

$$P(E)\, dE = -p(\theta)\, d\theta$$

$$= -p(\theta) \frac{d\theta}{dE}\, dE$$

(This is an example of the standard method for equating the distribution functions of different but related variables, in this case θ and E. The relationship in this case is equation (2.10) with the post-collision energy E_2 replaced by E. The negative sign in the preceding equation is necessary since as θ increases, E decreases, i.e. $d\theta$ and dE have opposite signs.)

From equation (2.10):
$$dE = -\frac{2AE_1}{(A+1)^2}\sin\theta\, d\theta \qquad (2.20)$$

Therefore:
$$P(E)\, dE = \frac{(A+1)^2}{4AE_1}\, dE$$

$$= \frac{dE}{(1-\alpha)E_1} \qquad (2.21)$$

The quantity $(1-\alpha)E_1$ is the range of possible post-collision energies, and equation (2.21) indicates that after scattering collisions neutrons are uniformly distributed throughout this range. Alternatively, neutron scattering may be said to be isotropic in post-collision energy in the range E_1 to αE_1. The average energy of neutrons after collision is $\frac{1}{2}(1+\alpha)E_1$, and the average loss of energy is $\frac{1}{2}(1-\alpha)E_1$.

Two important points emerge from these results. Firstly, the average loss of energy increases as α decreases, or as the mass number, A, of the scattering nucleus decreases, from which we conclude that light elements are more effective than heavy elements as far as neutron slowing down is concerned. Secondly, the average loss of energy is proportional to the pre-collision energy, and therefore decreases as neutrons lose energy. In order to have a quantity which characterizes the average loss of energy per collision and is at the same time independent of energy we will introduce the average decrease of (log energy) per scattering collision, $\overline{\log E_1 - \log E}$, which is given the symbol ξ (xi), and may be expressed as follows:

$$\xi = \int_{\alpha E_1}^{E_1}(\log E_1 - \log E)P(E)\, dE$$

$$= \frac{1}{(1-\alpha)E_1}\int_{\alpha E_1}^{E_1}\log\frac{E_1}{E}\, dE$$

By making the substitution $x = E/E_1$, we get:

$$\xi = \frac{1}{(1-\alpha)}\int_{1}^{\alpha}\log x\, dx$$

$$= 1 + \frac{\alpha}{1-\alpha}\log\alpha \qquad (2.22)$$

In terms of the mass number, A:

$$\xi = 1 - \frac{(A - 1)^2}{2A} \log \frac{A + 1}{A - 1} \qquad (2.23)$$

For values of A greater than about 10 the equation for ξ can be closely approximated by the simpler expression:

$$\xi = \frac{2}{A + 2/3} \qquad (2.24)$$

The quantity ξ, often called the logarithmic energy decrement, characterizes the slowing down power of an element for neutrons of any energy, and in the form given by equation (2.24) it is seen quite clearly that the most effective elements for slowing down neutrons are those of low mass number.

Frequently, slowing down takes place in a mixture of elements or a compound, and in this case the average value of the logarithmic energy decrement $\bar{\xi}$ for such a mixture is found as the weighted average of the values of ξ for each of the constituents of the mixture. The weighting factors are, logically, the probabilities of scattering in each of the constituents, which are proportional to the numbers of atoms of each constituent per cm³ of mixture multiplied by their scattering cross-sections. Thus $\bar{\xi}$ is calculated from the equation:

$$\bar{\xi} = \frac{\sum_i N_i \sigma_{si} \xi_i}{\sum_i N_i \sigma_{si}} \qquad (2.25)$$

Where N_i is the number of atoms per unit volume of the mixture of each element i whose microscopic scattering cross-section is σ_{si} and logarithmic energy decrement is ξ_i. The summation is carried out for all the constituents of the mixture.

In the case which is very common in nuclear engineering in which a heavy element such as uranium is mixed with a light element whose function is to slow down neutrons, the slowing down effect of the heavy element can often be neglected (provided the neutron energy is below the inelastic scattering threshold of that element), and for such a mixture the value of $\bar{\xi}$ is nearly the same as ξ for the light element.

In concluding this section we may mention that although neutron scattering in the C system is isotropic, and the average angle of scattering, θ, in the C system is 90°, neutron scattering is not isotropic in the L system. It can be shown that the average value of $\cos \psi$, where ψ is the angle of scattering in the L system, is given by:

$$\overline{\cos \psi} = \frac{2}{3A} \qquad (2.26)$$

where A is the mass number of the scattering nucleus. The inference from this result is that neutron scattering in the L system is preferentially in the forward direction since $\overline{\cos \psi}$ as given by equation (2.26) is always positive. As the mass of the scattering nucleus increases, $\overline{\cos \psi}$ approaches zero and the average angle of scattering approaches 90°. Thus for heavy elements it may be said that neutron scattering is nearly isotropic in the L system.

2.11 Thermal neutrons

In analysing elastic scattering collisions we have assumed that the nuclei of the medium are at rest since their speeds are much less than that of the neutrons. As the neutrons slow down their energies are reduced to values comparable with the energies of the nuclei, and a point is reached when scattering collisions no longer progressively reduce neutron energies. At this point neutrons have reached thermal equilibrium with the medium in which they are being scattered, and are known as thermal neutrons. The scattering of thermal neutrons has on the average no effect on their energy, although individual neutrons are at different energies and as a result of individual scattering collisions they may suffer either a slight increase or decrease of energy. The important point is that their average energy remains constant.

The situation is analogous to gas molecules in a rigid container at a constant temperature. The molecules are moving to and fro, colliding with each other and the walls of the container and gaining and losing energy, but the total energy (or average energy) of all the molecules remains constant. The distribution in terms of speed (or energy) of gas molecules is given by the Maxwell–Boltzmann distribution, and this function can be applied to thermal neutrons provided that their rate of absorption is small compared with their rate of scattering, that is provided $\sigma_a \ll \sigma_s$.

The Maxwell–Boltzmann distribution in terms of neutron speed, v, is:

$$n(v)\, dv = 4\pi N \left(\frac{m}{2\pi kT} \right)^{3/2} e^{-mv^2/2kT}\, v^2\, dv \qquad (2.27)$$

where $n(v)\, dv$ is the number of neutrons per unit volume whose speeds are in the range v to $v + dv$; N is the total number of neutrons per unit volume; m is the neutron mass, kg; T is the absolute temperature of the scattering medium, K; and k is the Boltzmann constant, $1 \cdot 38 \times 10^{-23}$ joule/particle K.

The upper curve of Figure 2.14 shows the shape of the Maxwell–Boltzmann distribution for neutrons in a medium at 20°C.

The most probable neutron speed, corresponding to the maximum point on this curve in Figure 2.14, is obtained by differentiating the distribution function with respect to v and equating to zero. The result is:

$$v_{MP} = \sqrt{\frac{2kT}{m}} \tag{2.28}$$

and the corresponding energy is $\frac{1}{2}mv_{MP}^2 = kT$.

Example: Calculate the most probable speed and the corresponding energy of thermal neutrons at 20°C.

$$v_{MP} = \sqrt{\frac{2 \times 1 \cdot 38 \times 10^{-23} \times 293}{1 \times 1 \cdot 66 \times 10^{-27}}} = 2200 \text{ m/s}$$

$$E = \frac{1 \cdot 38 \times 10^{-23} \times 293}{1 \cdot 602 \times 10^{-19}}$$

$$= 0 \cdot 0253 \text{ eV}$$

These values of the speed and kinetic energy are taken to characterize thermal neutrons in a medium at 20°C.

The average speed of thermal neutrons is determined from the distribution function by the standard statistical method:

$$\bar{v} = \frac{\int_0^\infty v n(v) \, dv}{\int_0^\infty n(v) \, dv} \tag{2.29}$$

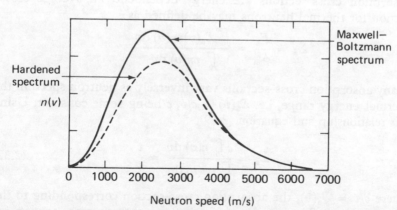

Figure 2.14. The Maxwell–Boltzmann distribution for thermal neutrons

The integration is carried out from zero to infinity to include all neutron speeds, although negligibly few neutrons have speeds greater than three or four times the most probable speed. The denominator in the preceding equation is equal to N, therefore, using equation (2.27):

$$\bar{v} = 4\pi \left(\frac{m}{2\pi kT}\right)^{3/2} \int_0^\infty v^3 \, e^{-mv^2/2kT} \, \mathrm{d}v$$

$$= \sqrt{\frac{8kT}{\pi m}} = \sqrt{\frac{4}{\pi}} \; v_{\mathrm{MP}} = 1 \cdot 128 \, v_{\mathrm{MP}} \qquad (2.30)$$

2.12 Thermal neutron flux, absorption rate and absorption cross-section

The thermal neutron flux is given by the equation:

$$\phi_{\mathrm{th}} = \int_0^{\sim 3v_{\mathrm{MP}}} \phi(v) \, \mathrm{d}v = \int_0^{\sim 3v_{\mathrm{MP}}} vn(v) \, \mathrm{d}v$$

If the Maxwell–Boltzmann distribution is used for $n(v) \, \mathrm{d}v$ the integration may be carried out from zero to infinity because, as already pointed out, there are very few neutrons in the Maxwell–Boltzmann distribution with speeds greater than $3v_{\mathrm{MP}}$. Thus:

$$\phi_{\mathrm{th}} = \int_0^\infty vn(v) \, \mathrm{d}v$$

The rate of absorption of thermal neutrons per cm^3, F_{a}, is:

$$F_{\mathrm{a}} = \int_0^\infty \Sigma_{\mathrm{a}}(v)\phi(v) \, \mathrm{d}v = \int_0^\infty \Sigma_{\mathrm{a}}(v)vn(v) \, \mathrm{d}v \qquad (2.31)$$

Since thermal neutrons have a spectrum of speeds or energies, and absorption cross-sections are energy dependent, an average cross-section for thermal neutrons may be defined as:

$$\bar{\Sigma}_{\mathrm{a}} = \frac{F_{\mathrm{a}}}{\phi_{\mathrm{th}}} = \frac{\int_0^\infty \Sigma_{\mathrm{a}}(v)vn(v) \, \mathrm{d}v}{\int_0^\infty vn(v) \, \mathrm{d}v}$$

Many absorption cross-sections vary inversely as neutron speed in the thermal energy range, i.e. $\Sigma_{\mathrm{a}}(v) = c/v$, c being some constant. Using this relationship and equation (2.29):

$$\bar{\Sigma}_{\mathrm{a}} = \frac{c \int_0^\infty n(v) \, \mathrm{d}v}{\int_0^\infty vn(v) \, \mathrm{d}v} = \frac{c}{\bar{v}} \qquad (2.32)$$

where $c/\bar{v} = \Sigma_{\mathrm{a}}(\bar{v})$, the absorption cross-section corresponding to the average speed. The conclusion is that the average cross-section for thermal neutrons in a $1/v$ absorber is equal to the cross-section corresponding to the average speed:

$$\bar{\Sigma}_{\mathrm{a}} = \Sigma_{\mathrm{a}}(\bar{v}) \qquad (2.33)$$

Absorption cross-sections are usually tabulated for neutrons with a speed of 2200 m/s, this being the most probable speed of thermal neutrons at 20°C, and it is more convenient to express the average cross-section in terms of this tabulated value, $\Sigma_a(2200)$. Recalling that $\bar{v} = \sqrt{(4/\pi)}v_{MP}$, it follows that:

$$\Sigma_a(\bar{v}) = \sqrt{\frac{\pi}{4}}\, \Sigma_a(v_{MP})$$

and for thermal neutrons at 20°C:

$$\Sigma_a(20°C) = \sqrt{\frac{\pi}{4}}\, \Sigma_a(2200) = 0.8862\, \Sigma_a(2200) \qquad (2.34)$$

The average cross-section for thermal neutrons at 20°C is $\sqrt{(\pi/4)}$ times the tabulated (2200 m/s) cross-section.

As can be seen from equations (2.28) and (2.30), the most probable and average neutron speeds are temperature dependent, and in fact vary as the square root of the absolute temperature. It follows that absorption cross-sections vary with temperature as follows:

$$\Sigma_a(T) \propto \frac{1}{\sqrt{T}}$$

or

$$\Sigma_a(T) = \Sigma_a(20°C)\sqrt{\frac{293}{T}}$$

The complete expression for the average cross-section of thermal neutrons in a medium at temperature T in terms of the tabulated value is:

$$\bar{\Sigma}_a(T) = 0.8862\sqrt{\frac{293}{T}}\, \Sigma_a(2200) \qquad (2.35)$$

The absorption rate, F_a, given by equation (2.31) can be expressed in a slightly different way by replacing $\Sigma_a(v)v$ in the integral by $\Sigma_a(2200)v_0$, where v_0 is 2200 m/s. The resulting expression for F_a is:

$$F_a = \Sigma_a(2200)v_0 N$$

$$= \Sigma_a(2200)\phi_0 \qquad (2.36)$$

ϕ_0 is called the 2200 m/s flux. It is a somewhat artificial concept as thermal neutrons do not all have a speed of 2200 m/s, however the significance of this artificial flux is that it is the flux that would be determined by measuring the thermal neutron absorption rate in a $1/v$ absorber and dividing it by the tabulated absorption cross-section. The calculation of absorption rates in a reactor is also simplified using the 2200 m/s flux as, regardless of the temperature of the reactor, the absorption rate is given by equation (2.36).

The results of this section apply to absorption rates in materials whose absorption cross-sections vary inversely as the neutron speed. In the case of some isotopes this relationship is not exact and a non-$(1/v)$ factor, g, is introduced to enable the absorption rate to be expressed as:

$$F_a = g\Sigma_a(2200)\phi_0 \tag{2.37}$$

Values of g for important reactor materials are tabulated in nuclear engineering literature, however a detailed treatment of this subject is beyond the scope of this book, and it will be assumed in all our calculations that the value of the factor g is 1 for all isotopes of interest.

In applying the Maxwell–Boltzmann distribution to describe the thermal neutron spectrum it has been assumed that neutron absorption is small compared with scattering. If the absorption rate is appreciable, the thermal neutron spectrum is distorted, and this distortion is most pronounced at the low energy end of the spectrum where, if $\sigma_a \propto 1/v$, the absorption is greatest. The distortion is shown in Figure 2.14 by the lower curve, and it can be seen that the effect of the distortion is to shift the peak of the spectrum to a slightly higher speed, although the general shape of the spectrum is not much altered. This effect is known as hardening of the spectrum.

Although the hardened spectrum is no longer precisely Maxwellian, it is often convenient to assume that it is, although shifted to a higher temperature as if the temperature of the medium were higher. The neutron temperature is defined as the temperature which, when used in the Maxwell–Boltzmann distribution, gives the best fit to the actual neutron spectrum in the absorbing medium. This method of representing the thermal neutron spectrum in an absorbing medium is only valid if the concentration of the absorber is small.

2.13 Reactor power, rating, fuel consumption and burnup

The calculation of the thermal power output of a reactor involves a knowledge of the mass of fissile material, its fission cross-section and the neutron flux. In the following example, which concerns a reactor fuelled with natural uranium, it will be assumed that all fission takes place in the ^{235}U and is caused by thermal neutrons. These assumptions, as we shall see in the following chapters, are not exactly true, however in many cases they are sufficiently accurate to give a good idea of the magnitude of the quantities involved.

Consider a reactor fuelled with 100 tonnes of natural uranium (1 tonne = 1000 kg), in which the average 2200 m/s flux is 10^{13} neutrons/cm^2 s. The 2200 m/s cross-sections of ^{235}U are:

$$\sigma_f = 579 \text{ barns}, \qquad \sigma_c = 101 \text{ barns}$$

The total number of ^{235}U atoms in the reactor

$$= \frac{10^5 \times 0.007\ 15 \times 6.023 \times 10^{26}}{238.05}$$

$$= 1.81 \times 10^{27} \text{ atoms}$$

Using equation (2.14), the rate of fission in the reactor

$$= 1.81 \times 10^{27} \times 579 \times 10^{-24} \times 10^{13}$$

$$= 1.05 \times 10^{19} \text{ fissions/s}$$

The rate of energy release or thermal power of the reactor

$$= 1.05 \times 10^{19} \times 200 \text{ MeV/s}$$

$$= \frac{1.05 \times 10^{19} \times 200 \times 1.602}{10^{19}} \text{ MW}$$

$$= 336 \text{ MW}$$

The rating of the reactor is the thermal power per unit mass of fuel and in this case is 3.36 MW/tonne of uranium. The rate of consumption of ^{235}U by fission is equal to the rate of fission, namely 1.05×10^{19} atoms/s. Expressed as a mass this is:

$$\frac{1.05 \times 10^{19} \times 235.04 \times 60 \times 60 \times 24}{6.023 \times 10^{26}} = 0.353 \text{ kg/day}$$
$$\text{(or 353 grams/day)}$$

A useful figure to keep in mind that is shown by this result is that the complete fissioning of 1 gram of ^{235}U releases about 1 MWd of thermal energy. It requires about 3 tonnes of coal to release the same amount of energy in a combustion process.

The total rate of consumption of ^{235}U is greater than the fission rate due to the fact that ^{235}U is also used up when it captures neutrons to form ^{236}U. The total consumption rate of ^{235}U in this example is:

$$\frac{\sigma_c + \sigma_f}{\sigma_f} \times 0.353 = 0.415 \text{ kg/day (or 415 grams/day)}$$

The burnup of nuclear fuel is a measure of the total amount of energy released by fission per unit mass of fuel over a period of time. In this example, if the reactor operates at steady and continuous power for one year the burnup is $3.36 \times 365 = 1226$ MWd/tonne of uranium. The fraction of ^{235}U consumed, both by fission and neutron capture, in one year's continuous operation is:

$$\frac{0.415 \times 365 \times 6.023 \times 10^{26}}{235.04 \times 1.81 \times 10^{27}} = 0.214$$

(Note that if the reactor operates at constant power for a long period, and if all fission occurs in ^{235}U, then the neutron flux does not remain constant, but increases to compensate for the decreasing amount of ^{235}U in the reactor.)

It should be emphasized, however, that these results depend on the assumption that all fission takes place in ^{235}U. In any reactor which contains ^{238}U, as we shall see in the next chapter, neutron capture in ^{238}U leads to the production of the fissile isotope ^{239}Pu. Once this isotope is created in the reactor, a fraction of the fission occurs in it and the amount of ^{235}U used is less than the figure calculated above, provided the reactor continues to operate at constant power.

Chapter 3

The chain reaction and principles of nuclear reactors

It is now appropriate to consider the necessary requirements for a chain reaction, and also some other topics that are of importance in the development of nuclear energy for peaceful purposes.

The achievement of a system in which a controlled, self-sustaining fission chain reaction takes place is the first requirement, as this is the way in which the energy of fission can be released, also in a controlled way, and put to good use. The system in which the chain reaction takes place is called a nuclear reactor, and there are many possible types of reactor, depending on the materials of construction and the energy of the neutrons which cause fission. In this chapter we will consider qualitatively the conditions necessary for a chain reaction and will identify a number of different possible reactor types. Some types of reactor require enriched uranium to achieve a chain reaction, and the processes for enrichment are briefly described.

The complete utilization of the world's resources of uranium to provide energy is an important aspect of nuclear power, and the discussion of this topic leads to a description of the types of reactor and the fuel cycles that will enable not only uranium, but also thorium to be used as a long-term source of energy.

Finally, the role which nuclear power can play in supplying a world in which energy demands are increasing and resources are being rapidly depleted will be discussed.

3.1 The chain reaction

The condition that is necessary for a stable, self-sustaining chain reaction is that exactly one of the neutrons produced in one fission event proceeds to cause a second fission from which one neutron goes on to cause a third fission, and so on. In such a reaction the neutron density and fission rate remain constant. This condition can be expressed by means of a multiplication factor, k, which is defined as the ratio of the number of neutrons in one generation to the number of neutrons in the preceding generation. When this factor is exactly one the condition for a stable chain reaction is satisfied and the reactor is said to be critical. If this factor is greater than one the reactor is supercritical and a divergent chain reaction exists in which the neutron density and fission rate increase, possibly at an explosive rate as in an atomic bomb. If the multiplication factor is less than one the reactor is subcritical and the chain reaction decreases and eventually dies out.

A nuclear reactor is an assembly of many components of which at this stage we need mention a few of the most important. The most important component of any reactor is the fuel in which fission takes place and energy, in the form of heat, is released. At the present time uranium is the most widely used nuclear fuel, although the isotope ^{239}Pu is becoming increasingly important. In many reactors a light element is included for the specific purpose of slowing down neutrons from fission to thermal energy at which they are most effective for causing further fission. This material is called the moderator and we shall discuss moderators and their characteristics later in this chapter. Finally cladding materials are required to contain and support the fuel and prevent the release of radioactive fission products, and in the case of all reactors, except those operating at very low power, a coolant is required which is circulated through the core to transport the energy released in the fuel by fission to external heat exchangers.

Returning to the conditions for a critical reactor we can say that the rate at which neutrons are used up or lost in a reactor must be exactly equal to the rate at which they are produced by fission. Neutrons are used up or lost in two ways, either they leak out of the reactor or they are absorbed within the reactor. Absorption within the reactor includes fission and capture in the fuel and capture in the other materials of the reactor such as the moderator, coolant, cladding and control rods. The relative rates at which these processes take place depend on the size and composition of the reactor. It is the purpose of reactor theory to analyse the various processes in a reactor, determine their rates and calculate for a given composition and size of core the value of the multiplication factor. Alternatively the calculation may be directed towards finding the critical size or composition of a reactor for a given value of the multiplication factor.

As an introduction to this type of problem let us consider the possibility of a chain reaction in an infinitely large system of pure natural uranium which consists of 99·285 per cent ^{238}U and 0·715 per cent ^{235}U. (The ratio of ^{238}U atoms to ^{235}U atoms is approximately 139 to 1.) The purpose of considering an infinite system is to avoid for the present time the question of neutron leakage as neutrons cannot leak out of an infinite system.

Neutrons produced by fission in the uranium have an average energy of 2 MeV and at this energy the significant cross-sections are: ^{238}U; $\sigma_i = 2.87$ barns, $\sigma_f = 0.6$ barns; ^{235}U; $\sigma_i = 2.3$ barns, $\sigma_f = 1.3$ barns. (The values of the other cross-sections are negligibly small.) Bearing in mind the greater amount of ^{238}U, it is evident that inelastic scattering in ^{238}U will be the dominant process for 2 MeV neutrons, which will rapidly lose energy and fall below the threshold for fission in ^{238}U (see Figure 2.12). There may be some fission in ^{238}U caused by neutrons of energy greater than 1 MeV, but it will not be sufficient to establish a chain reaction. At about 0.3 MeV the significant cross-sections are: ^{238}U; $\sigma_i = 0.5$ barns; ^{235}U; $\sigma_i = 0.7$ barns, $\sigma_f = 1.3$ barns. Bearing in mind the ^{238}U to ^{235}U ratio, it is evident that inelastic scattering in ^{238}U continues to be the most probable type of interaction and the neutron energy falls to about 1000 eV, the upper limit of the resonance region of ^{238}U (see Figure 2.9). Below this energy the capture cross-section of ^{238}U rises to isolated values or resonances which are much higher than the scattering cross-section. The fission cross-section of ^{235}U has similar resonances between 100 eV and 1.0 eV, but they are not so high as the ^{238}U resonances. In this energy range, therefore, neutron capture in ^{238}U becomes the dominant process to such an extent that practically all neutrons are captured in ^{238}U resonances and negligibly few cause fission in ^{235}U. A chain reaction is thus impossible in pure natural uranium.

Let us now consider the possibility of a chain reaction in an infinite system of enriched uranium with 50 per cent ^{235}U and 50 per cent ^{238}U. Neutrons with an energy of 2 MeV have, referring to the cross-section values already quoted, about a 27 per cent chance of causing fission and a 73 per cent chance of being inelastically scattered. This is not sufficient to establish a chain reaction at 2 MeV. At lower energies the values of σ_i for both ^{235}U and ^{238}U decrease and the value of σ_f for ^{235}U increases so that at 0.3 MeV, using the previous figures, fission occurs in just about 50 per cent of interactions and a chain reaction is possible. The conclusion to be drawn from this very qualitative argument is that in a system of 50 per cent ^{235}U and 50 per cent ^{238}U a chain reaction can be established by neutrons in the energy range 0.3 to 2 MeV. This is the basis of a fast reactor, the word fast indicating that the fission causing neutrons have high energy.

Turning our attention again to an infinite system of natural uranium the question arises as to whether there is any neutron energy at which fission in ^{235}U would be a sufficiently probable reaction to establish a chain reaction. Considering thermal neutrons whose energy is taken as 0.025 eV, the values of the relevant cross-sections are: ^{238}U; $\sigma_c = 2.72$ barns; ^{235}U; $\sigma_c = 101$ barns, $\sigma_f = 579$ barns. (Scattering need not be considered as the scattering of thermal neutrons does not, on average, affect their energy.) Using these figures the fraction of thermal neutrons absorbed in natural uranium which cause fission is:

$$\frac{579}{579 + 101 + 2 \cdot 72 \times 139} = 0 \cdot 547$$

If $2 \cdot 42$ neutrons are produced per fission and $54 \cdot 7$ per cent of these cause fission then the multiplication factor is $2 \cdot 42 \times 0 \cdot 547 = 1 \cdot 32$, and a chain reaction is established. This is of course a highly simplified argument and has ignored several important aspects such as neutron capture in other components of the reactor and neutron leakage in a reactor of finite size, however the fact remains that there is a possibility of establishing a chain reaction in natural uranium with thermal neutrons.

The problem now arises as to how to slow down fission neutrons to thermal energy and at the same time ensure that as few as possible are captured in the ^{238}U resonances during the process. This is achieved by combining the uranium fuel with a material, called the moderator, which is effective at slowing down neutrons. The moderator should be an element of low mass number (or a compound containing such an element), it should have a high scattering cross-section if possible, and in order that it does not capture neutrons (which would reduce the possibility of a chain reaction) it should have a low capture cross-section. Furthermore, in order that scattering should be the dominant process at all energies, including the ^{238}U resonance region, the quantity of moderator should be much greater than the quantity of fuel. A reactor in which fission is caused predominantly by thermal neutrons is called a thermal reactor.

The choice of the moderator for a thermal reactor is governed by the considerations mentioned in the last paragraph, and there are three materials which are suitable and widely used, namely water, heavy

Table 3.1. *Properties of some important moderators*

	Water H_2O	Heavy water D_2O	Graphite	Beryllium
σ_a (barns)	0·66	0·001	0·0045	0·0092
σ_s (barns)	~50	10·6	4·7	6·1
ξ or $\bar{\xi}$	0·920	0·509	0·158	0·209
No. of collisions to thermalize fission neutrons	20	36	115	87
Moderating power $\xi\sigma_s$	46	5·4	0·74	1·27
Moderating ratio $\xi\sigma_s/\sigma_a$	1390	5400	165	139

water and carbon in the form of graphite. Beryllium, beryllium oxide and certain hydrocarbon compounds might also be suitable, but they each have certain disadvantages which have prevented their large scale use as moderators. The important properties of some of these materials are listed in Table 3.1.

Each of the moderators listed in Table 3.1 has certain advantages and disadvantages which makes the choice of one of them as the best possible moderator more or less impossible. Water is cheap and has excellent slowing-down properties and a high scattering cross-section, however its capture cross-section is rather high and enriched uranium is required as fuel in a water-moderated reactor. Heavy water has good slowing down properties and a very low capture cross-section. It is, however, very expensive to separate heavy water from ordinary light water, and this economic factor is a disadvantage in the use of heavy water as the moderator in large reactors. Graphite is fairly cheap and has a low capture cross-section, but its mass number is rather too high for it to be regarded as an ideal moderator; it is nevertheless extensively used, particularly in British power reactors. Beryllium is rather expensive and toxic and has not been used as a moderator on a large scale.

3.2 The neutron cycle in a thermal reactor

Accurate calculations of the multiplication factor for any reactor must take full account of all the processes involving neutrons between the time when they are produced by fission and the time when they eventually disappear either by absorption or by leakage from the reactor. The procedure is simplified if we consider various parts of the neutrons' lifetime separately, and we will apply this method to a thermal reactor of finite size fuelled with natural or enriched uranium. The neutron cycle for such a reactor is illustrated in Figure 3.1.

Consider n neutrons at an average energy of 2 MeV produced by thermal fission in ^{235}U. Before these neutrons slow down below 1 MeV there is a possibility that a few cause fission, referred to as fast fission, in ^{238}U. The fast fission factor, ϵ, is defined as:

The number of neutrons slowing down below 1 MeV per neutron produced by thermal fission.

Now $n\epsilon$ neutrons are slowing down below 1 MeV and continue to slow down as a result principally of elastic scattering collisions with the moderator. During the slowing down process some neutrons leak out of the reactor, and some neutrons are captured in the ^{238}U resonances. The resonance escape probability, p, is defined as:

The fraction of neutrons which escape capture in the ^{238}U resonances during slowing down in the reactor.

The fast non-leakage probability, P_{NLf}, is defined as:

The fraction of neutrons which do not leak out of the reactor during slowing down.

The number of neutrons which slow down and become thermal is $n\epsilon p P_{NLf}$.

Once the slowing down process is complete and neutrons are thermalized they will continue to diffuse in the reactor until they are absorbed or leak out. Only a fraction of the absorption of neutrons is in the fuel, and some neutrons are captured in the moderator, coolant or structural

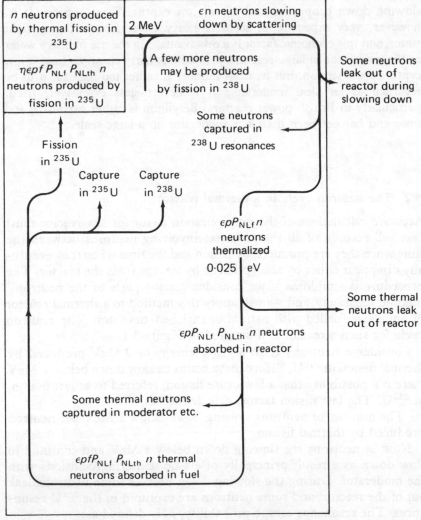

Figure 3.1. The neutron cycle in a thermal reactor

materials of the core. The thermal non-leakage probability, P_{NLth}, is defined as:

The fraction of thermal neutrons which do not leak out of the reactor.

The thermal utilization factor, f, is defined as:

The fraction of thermal neutrons absorbed in the reactor which are absorbed in the fuel.

It is evident that the number of neutrons which is absorbed in the fuel, is $n\epsilon pfP_{NLf}P_{NLth}$. Not all these neutrons cause fission, in fact in natural uranium our earlier calculation indicates that about 55 per cent of thermal neutrons absorbed in the fuel cause fission, and that the number of neutrons produced is about 1·33 per neutron absorbed in the fuel. This quantity, the average number of fission neutrons produced per neutron absorbed in the fuel, has the symbol η and is usually just called eta.

Finally, therefore, at the beginning of the second generation $n\epsilon pf\eta P_{NLf}P_{NLth}$ fission neutrons are produced from n fission neutrons at the beginning of the first generation. From our definition of the multiplication factor, which is called more explicitly for a finite reactor the effective multiplication factor, k_{eff}, we can write:

$$k_{eff} = \epsilon pf\eta P_{NLf}P_{NLth} \tag{3.1}$$

It is often convenient to consider an infinitely large reactor, as this enables us to neglect neutron diffusion and leakage, and $P_{NLf} = P_{NLth} = 1$. The multiplication factor is then referred to as the infinite multiplication factor, k_∞, and is expressed by the important Four Factor Formula:

$$k_\infty = \epsilon pf\eta \tag{3.2}$$

The relationship between the two multiplication factors is:

$$k_{eff} = k_\infty P_{NLf}P_{NLth} \tag{3.3}$$

The calculation of k_{eff} or k_∞ can be carried out by calculating separately the factors on the right-hand side of equation (3.1) or (3.2), and some of the elementary theory underlying these calculations will be dealt with in the next chapter.

Earlier in this chapter, it was pointed out that for certain types of reactor enriched uranium is necessary to achieve criticality. The most important example is the pressurized water reactor which requires slightly enriched uranium with 2 to 3 per cent of ^{235}U. Some current British graphite moderated reactors use fuel of a similar composition. Uranium fuelled fast reactors (of which very few have been built) require highly enriched uranium with 25 to 50 per cent of ^{235}U.

The process of enriching uranium involves a partial separation of the ^{235}U and ^{238}U so that the product has a higher concentration of ^{235}U than does natural uranium. The waste, known as the tails, has a lower concentration. Two processes are available for uranium enrichment on a commercial scale. In both of them, natural uranium is converted to the gaseous compound uranium hexafluoride UF_6 and the two isotopes of uranium produce two gases of slightly different density, the $^{238}UF_6$ being slightly more dense than the $^{235}UF_6$. Both processes make use of this slight difference in density to achieve separation of the isotopes. In the gaseous diffusion process the UF_6 gas diffuses through a series of semi-permeable membranes, and in the centrifuge process the UF_6 gas is spun at high rotational speed in a centrifuge. Further description of these processes is postponed to Chapter 12.

3.3   Conversion and breeding

One important point to emerge from earlier sections of this chapter is that in thermal reactors fuelled with uranium, either natural or enriched, practically all the fission occurs in ^{235}U. In fast reactors, which contain no moderator and in which neutron energies are much higher than in thermal reactors, ^{238}U fission occurs to a small extent, but even in this type of reactor it is ^{235}U fission which predominates and sustains the chain reaction. It is nearly correct to say, therefore, that only the ^{235}U in natural uranium contributes energy directly from its own fission.

Although ^{238}U cannot itself be used as the fuel in a nuclear reactor, it does have a vital role to play as an isotope from which new fissile fuel can be created. In a uranium fuelled reactor a significant fraction of the neutrons produced by fission, possibly 30–40 per cent, are captured in ^{238}U and produce ^{239}U by an (n, γ) reaction. ^{239}U is the start of a radioactive decay chain which produces neptunium 239, also radioactive and plutonium 239, which has a very long half-life and can almost be regarded as a stable isotope. The processes involved are:

$$^{238}U\,(n,\,\gamma)\,^{239}U\xrightarrow[T_{\frac{1}{2}}\,=\,23\ \text{min}]{\beta^-}\,^{239}Np\xrightarrow[T_{\frac{1}{2}}\,=\,2\cdot3\ \text{days}]{\beta^-}$$
$$^{239}Pu\;(T_{\frac{1}{2}}\,=\,25\,000\ \text{y}).$$

Plutonium 239, as already pointed out, is fissile, its characteristics as far as fission is concerned are similar to those of ^{235}U, and it can be used as the fuel in both fast and thermal reactors.

Another isotope which has characteristics similar to ^{238}U is ^{232}Th, the only naturally occurring isotope of the element thorium. This isotope can only undergo fission with neutrons of energy greater than about 1·4 MeV, so it cannot sustain a chain reaction and be used directly as a nuclear fuel. However, as a result of neutron capture in ^{232}Th the following processes take place:

$$^{232}\text{Th} \ (n, \gamma) \ ^{233}\text{Th} \ \xrightarrow[\ T_{\frac{1}{2}} = 22 \ \text{min}\ ]{\beta^-} \ ^{233}\text{Pa} \ \xrightarrow[\ T_{\frac{1}{2}} = 27 \cdot 4 \ \text{days}\ ]{\beta^-}$$

$$^{233}\text{U} \ (T_{\frac{1}{2}} = 1 \cdot 6 \times 10^5 \ \text{y}).$$

Uranium 233 is fissile with neutrons of all energies and like ^{235}U and ^{239}Pu it can be used as the fuel for nuclear reactors.

The importance of ^{238}U and ^{232}Th lies in their ability to act as fertile materials from which, as a result of neutron capture in nuclear reactors, the fissile isotopes ^{239}Pu and ^{233}U are produced. This process is known as breeeding or conversion, and it provides the method whereby ^{238}U and ^{232}Th can be used as sources of energy through fission.

An important characteristic of a nuclear reactor, particularly one which is designed to produce new fissile material by one of the processes just described as well as power, is the ratio of the rate at which new fissile atoms are produced to the rate at which existing fissile atoms are used up. This ratio is the breeding ratio B and is defined as:

The number of new fissile atoms produced in a reactor per atom of existing fissile fuel consumed by fission and neutron capture.

If the breeding ratio is exactly 1 then the quantity of fissile fuel remains constant, if it is greater than 1 the quantity of fissile fuel increases, and if it is less than 1 the quantity of fissile fuel decreases. In order to utilize all the world's resources of ^{238}U and ^{232}Th it is essential that some, though not necessarily all, of the world's reactors are designed so that their breeding ratios are greater than 1.

The following simplified argument illustrates this point. Consider a reactor (or a number of reactors) whose breeding ratio is B, fuelled with natural uranium, the total mass of ^{235}U being M tonnes. If all the ^{235}U is used up, the quantity of ^{238}U converted to ^{239}Pu is approximately BM tonnes. If this plutonium is used to provide the second charge of fuel for the reactors, and if the value of B is the same for a plutonium fuelled reactor as for a uranium fuelled reactor (which is not strictly true as we shall see), then the use of BM tonnes of ^{239}Pu results in another B^2M tonnes of ^{238}U being converted to ^{239}Pu, and so on. The total amount of uranium used up would be $M + BM + B^2M + \ldots$. If $B < 1$, then this is equal to $M/(1 - B)$. If this result were exact, which it is not, the value of B necessary to use up all the natural uranium originally in the reactors would be about $0 \cdot 993$, or for all practical purposes 1. In fact it is necessary that the value of B should be slightly greater than 1 as the foregoing argument neglects the inevitable losses of plutonium and uranium during chemical processing, separation of fission products and manufacture of fuel elements.

If the value of B is much less than 1 then only a fraction of the available uranium is used up, for example if $B = 0 \cdot 75$, then theoretically

2·86 per cent of the natural uranium can be used, still a very small fraction which in practice would be even smaller.

The dependence of the breeding ratio on other reactor parameters can be deduced from the following argument. When a neutron is absorbed in an atom of fissile fuel, that atom is consumed (in the sense used in the definition of the breeding ratio) and η_f neutrons are produced, where η_f is the average number of neutrons produced per neutron absorbed in the fissile fuel. For a steady chain reaction in a reactor one of these neutrons must be absorbed in another atom of fissile fuel to keep the reaction going, and according to the definition of the breeding ratio, B neutrons must be captured in the fertile material. Some neutrons will inevitably be captured in non-fuel materials, and some will leak out of the reactor. The sum of these two processes is represented by $(C + L)$ neutrons per neutron absorbed in the fissile fuel. These processes are shown in Figure 3.2. It is clear that in order to maintain a steady chain reaction:

$$\eta_f = 1 + B + (C + L)$$

$$\text{or} \quad B = \eta_f - 1 - (C + L) \tag{3.4}$$

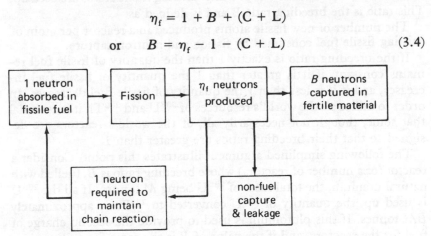

Figure 3.2. Neutron cycle to illustrate breeding

Clearly if B is to be greater than 1, η_f must exceed 2 by an amount that allows for the term $(C + L)$, whose value is likely to be about 0·2. Values of η_f for the three fissile isotopes for fission caused by thermal and high energy neutrons are given in Table 3.2.

It will be seen from Table 3.2 that, allowing for reasonable non-fuel neutron capture and leakage, only a ^{233}U fuelled thermal reactor can achieve a breeding ratio greater than 1. In fast reactors there is a considerable improvement, and reactors of this type are capable of giving breeding ratios greater than 1. The value of B in a fast reactor is actually greater than that given by equation (3.4) due to the effect of fast fission in ^{238}U or ^{232}Th, which is very slight in a thermal reactor but may be considerable in a fast reactor in which as much as 20 per cent of the fission may be in the ^{238}U (less in ^{232}Th).

Table 3.2. *Typical values of η_f for the fissile isotopes*

Isotope	Energy of neutrons causing fission	
	Thermal (0·025eV)	High energy (0·5 MeV)
^{235}U	2·07	2·35
^{239}Pu	2·15	2·90
^{233}U	2·29	2·40

3.4 Fuel cycles and breeder reactors

In view of what has been said about breeding and the characteristics of the fissile isotopes in thermal and fast reactors, it is possible to visualize certain types of reactor and their associated fuel cycles.

Figure 3.3 shows a thermal reactor fuelled initially with natural or enriched uranium; it is typical of the vast majority of reactors built up to the present. The breeding ratio of such a reactor is less than 1 so the amount of ^{239}Pu produced is less than the amount of ^{235}U used. When the reactor is refuelled, although the ^{239}Pu might be recycled with the depleted uranium (which contains much less than 0·715 per cent of ^{235}U), some additional fissile fuel is needed to make up the deficit. Up to 1980 practically all the plutonium produced in this way has been stockpiled for nuclear weapons and future reactors, so that up to the present nearly all nuclear reactors have been fuelled with natural or enriched uranium and stocks of plutonium and depleted uranium are steadily accumulating in several countries.

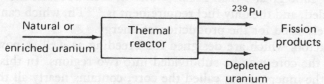

Figure 3.3. Thermal reactor fuelled with uranium

Future reactors will be fuelled with ^{239}Pu and depleted uranium, which can be regarded as pure ^{238}U, and in such reactors ^{238}U will be converted to ^{239}Pu, but only in fast reactors will the breeding ratio be greater than 1 and in such reactors more ^{239}Pu will be produced than is destroyed. Figure 3.4 shows the fuel cycle of a fast reactor fuelled with ^{239}Pu and ^{238}U in which the breeding ratio is greater than 1.

When the first fuel charge is unloaded from the reactor it contains more ^{239}Pu and less ^{238}U than it did when new, and it is contaminated with fission products which must be removed during reprocessing. Some ^{239}Pu is available for other uses such as fuelling thermal reactors, while the rest is recycled. The ^{238}U is recycled but there is less of this

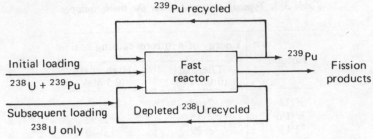

Figure 3.4.   Fuel cycle for a fast reactor fuelled with ^{239}Pu and ^{238}U

isotope than in the original charge, so an additional supply ^{238}U is required for the second and subsequent fuel loadings. The important point about this fuel cycle is that the reactor is kept going by fresh supplies of non-fissile ^{238}U, and in this way ^{238}U can be completely converted to ^{239}Pu and used as a source of energy.

If the reactor of Figure 3.4 had been a thermal reactor in which the breeding ratio is less than 1, it would require a continuous supply of ^{239}Pu as well as ^{238}U to keep it going. It is obvious that in a planned long-term nuclear power programme a combination of fast and thermal reactors will be able to utilize all the world's resources of uranium.

Similar ideas apply to reactors making use of the ^{232}Th–^{233}U breeding process. A reactor fuelled initially with either ^{235}U or ^{239}Pu as the fissile material, and ^{232}Th as the fertile material will produce ^{233}U. When sufficient ^{233}U is produced it can be used as the fuel charge in either a fast or thermal reactor with ^{232}Th, and in such a reactor a breeding ratio greater than 1 is possible. In such a reactor the ^{233}U can be recycled, and the only fuel requirement is ^{232}Th, which can thus be completely used for the production of energy.

In reactors which are designed for breeding as well as power production the core may be subdivided into two regions. In this type of reactor the inner region, called the core, contains nearly all the fissile material (^{239}Pu), and it is in this part of the reactor that most of the energy is released by fission. The outer region is called the blanket and contains the fertile material (^{238}U). At the start of the reactor's life there is very little fission in the blanket. However neutrons produced by fission in the core may diffuse into the blanket and be captured in ^{238}U to produce ^{239}Pu. As the operation of this reactor proceeds, fissile material builds up in the blanket and provision must be made to remove the energy released by fission from the blanket as well as from the core.

Another point worthy of mention is that in any reactor designed for breeding, neutron capture in the moderator, structural materials, etc., and neutron leakage from the core should be reduced to a minimum. This capture and leakage is the term $(C + L)$ in equation (3.4), and it is clear from this equation that any increase of the term $(C + L)$ reduces the possible value of the breeding ratio. As an example of this, it is generally true to say that the breeding ratio in a water-moderated reactor is less than in a graphite or heavy water-moderated reactor due to the rather high capture cross-section of water. In a reactor with a reasonably high value of the breeding ratio (say between 0·8 and 1·0) new fissile fuel is being produced almost as fast as it is being consumed, and the fuel in such a reactor can be used to a very high burnup, thus prolonging the periods between refuelling and reducing fuel costs.

In an expanding programme of nuclear power using breeder reactors to produce new fuel for later reactors, an important parameter is the doubling time, T_d, which is defined as the time required for the quantity of fissile fuel in a breeder reactor to double itself. Clearly, it can be interpreted as the time required for a breeder reactor to produce enough new fuel to provide the first fuel charge for another identical reactor, and this time will control the rate at which a breeder reactor programme can be expanded.

An expression for the doubling time can be derived as follows:-
The rating of a reactor is given by:

$$R = N_F \sigma_f \phi E \text{ watts/gram} \tag{3.5}$$

where N_F is the number of atoms per gramme of fissile fuel and E is the energy released per fission, joules.

If the mass of fissile fuel in the reactor is M_F grams, then the rate of consumption of fissile fuel is $M_F N_F \sigma_a \phi$ atoms/second, and the rate of production of new fissile fuel is $B M_F N_F \sigma_a \phi$ atoms/second. The net rate of production of new fissile fuel, i.e. the rate at which production exceeds consumption $= (B-1) M_F N_F \sigma_a \phi$ atoms/second. This rate multiplied by the doubling time must by definition be equal to the original number of atoms of fissile fuel $M_F N_F$. Thus:

$$T_d (B - 1) M_F N_F \sigma_a \phi = M_F N_F$$

Substituting for ϕ from equation (3.5) and rearranging, we get:

$$T_d = \frac{N_F \sigma_f E}{(B - 1) R \sigma_a} \tag{3.6}$$

Since σ_f, σ_a, N_F and E are all constant for any particular fissile isotope, equation (3.6) can be interpreted as stating that the doubling time is inversely proportional to both $(B - 1)$ and the reactor rating R. In order to reduce the doubling time it is therefore desirable to design a breeder reactor with as high a value of B as possible, and to operate the reactor at as high a rating as possible.

Doubling times are measured in years, rather than weeks or months. For example, if the following values for a ^{239}Pu fuelled fast breeder reactor are used, namely $N_F = 2 \cdot 52 \times 10^{21}$ atoms/gram, $E = 3 \cdot 36 \times 10^{-11}$ joules/fission, $\sigma_f = 1 \cdot 8$ barns and $\sigma_a = 2 \cdot 15$ barns, and if the breeding ratio is assumed to be $1 \cdot 2$ and the rating 500 MW/tonne of ^{239}Pu, then the doubling time can be calculated to be about 22 years.

The operation of fast breeder reactors at high ratings is also dictated by the fact that the mass of fuel in these reactors is much smaller than in thermal reactors, and if these reactors are to produce large quantities of energy for, say, electricity generation, then they must operate at high ratings. The high ratings characteristic of fast breeder reactors lead to certain engineering features which will be described in a later chapter.

3.5 Classification of reactor types

It is possible to summarize the several different types of nuclear reactors which have been developed in the world to date, classifying them according to the choice of fuel and other materials of construction, including the moderator.

There are different forms in which the uranium fuel can be used in the reactor. One possibility is pure uranium which is a dense and rather soft metal which undergoes a metallurgical phase change at 650°C, and this is the maximum operating temperature for metallic uranium to eliminate the possibility of distortion of the fuel resulting from the phase change. An alternative and much more common form in which uranium is used is as uranium dioxide (UO_2), a powder which can be fabricated into pellets which are assembled in stainless steel or zirconium alloy tubes to form fuel rods. Uranium dioxide has a very high melting point, about 2700°C, and reactors using this type of fuel can operate at higher fuel temperatures than those using metallic uranium. In the case of fast breeder reactors, as was shown in Sections 3.3 and 3.4, plutonium 239 is an important fuel. It is used in the form of plutonium oxide mixed with uranium dioxide to give a mixed oxide (MOX) fuel containing typically about 25 per cent of plutonium oxide.

As was pointed out at the end of Section 3.1, the three most important moderators, and the only ones in use in present thermal reactors, are water, heavy water and graphite. The saturation temperature/pressure properties of water and heavy water make it necessary for reactors moderated with these materials to operate at lower temperatures than graphite moderated reactors.

In the case of power reactors it is necessary to circulate a coolant through the reactor core to remove the energy released by fission and transfer this energy to the steam power cycle. The choice of coolant influences the reactor design and the temperature at which it operates. Either a liquid or a gas may be chosen, the most important to date being water, heavy water, liquid sodium and carbon dioxide, with helium a possible coolant for future gas cooled reactors.

A classification of types of nuclear reactor developed in the world to date is shown below in Table 3.3.

Table 3.3. Principal types of nuclear reactor

Fuel	Moderator	Coolant	Name	Country of origin
Uranium dioxide (slightly enriched)	Water	Water (non-boiling)	Pressurized Water Reactor (PWR)	USA
Uranium dioxide (slightly enriched)	Water	Water (boiling)	Boiling Water Reactor (BWR)	USA
Uranium (natural)	Graphite	Carbon dioxide	Gas Cooled Reactor (GCR)	UK
Uranium dioxide (slightly enriched)	Graphite	Carbon dioxide	Advanced Gas Cooled Reactor (AGR)	UK
Uranium dioxide (slightly enriched)	Graphite	Helium	High Temperature Gas Cooled Reactor (HTGR)	UK USA FDR
Uranium dioxide (natural)	Heavy water	Heavy water (non-boiling)	Pressurized Heavy Water Reactor (PHWR)	Canada
Uranium dioxide (slightly enriched)	Heavy water	Water (boiling)	Steam Generating Heavy Water Moderated Reactor (SGHWR)	UK
Uranium dioxide (slightly enriched)	Graphite	Water (boiling)	Graphite Moderated Boiling Water Reactor (RBMK)	USSR
Uranium dioxide + Plutonium oxide	None	Liquid sodium	Fast Breeder Reactor (FBR)	Various

3.6 Uranium as an energy resource

The role which uranium can play in contributing to the world's energy needs is potentially very great, and the development of nuclear power will only be justified in the long run if uranium and other fissile fuels are able to make a substantial contribution to energy supplies for many decades to come. This will depend on at least three factors: the extent of uranium resources which can be used, the way in which uranium is used in breeder and non-breeder reactors, and public acceptance of nuclear power.

Uranium, in the form of a mixture of oxides whose approximate formula is U_3O_8, is found quite widely in the earth's crust in very small concentrations. The largest reserves in non-Communist countries are in Canada, Australia, South Africa and the USA. The richest ores contain 0.05 to 0.2 per cent U_3O_8, and those which are at present being exploited produce low cost uranium at prices less than \$80/kg U. These ores will be able to satisfy the present (1987) annual consumption of 41 000 tonnes for at least 40 years. Much poorer ores in shales and granites containing less than 0.01 per cent U_3O_8 will then have to be exploited, and the price of uranium will then become much higher than the figure quoted above. The environmental impact of processing huge quantities of rock and shale to obtain relatively small quantities of uranium will be considerable. Uranium is also present in minute quantities in sea water, but it is doubtful if it will ever be economically justifiable to exploit this source.

Uranium resources are classified according to their production costs (\$/kg U) and the levels of confidence in the resource data. Reasonably assured resources (RAR) refer to known deposits of uranium ore that can be recovered by proven technology. Estimated additional resources (EAR) refer to quantities which are expected to exist as an extension to existing deposits (EAR-1) or in other areas of mineralization which have not been fully explored (EAR-2). Finally, speculative resources (SR) refer to deposits of uranium ores that are thought to exist on the basis of extrapolation and indirect evidence.

Table 3.4. Uranium resources
(World excluding Communist countries)

Cost of Production	Millions of tonnes of Uranium			
	RAR	EAR-1	EAR-2	SR
Up to \$80/kg U	1·67	0·92	0·60	4·16
\$80 to \$130/kg U	0·65	0·41	1·01	
\$130 to \$260/kg U	0·42	0·44	0·63	3·04
Total up to \$260/kg U	2·74	1·77	2·24	7·20

Source: OECD

The total of almost 14 million tonnes does not include the reserves in Communist countries which are estimated to be between 3·3 and 8·4 million tonnes.

The total of 3·65 million tonnes of uranium which includes RAR and EAR-1 up to costs of \$130/kg U are regarded as the resource which the nuclear industry can be confident will be available for the thermal reactor programme of the next few decades.

The energy value of these reserves depends on the type of reactor in which the uranium is used. Two possible examples are:

(a) thermal non-breeder reactors in which a total of 2 per cent of uranium is used, and

(b) fast breeder reactors in which all uranium is used.

The energy values and coal equivalents of uranium for these two cases are:

(a) 1 kg of uranium has an energy value of $1 \cdot 2 \times 10^9$ kJ and is equivalent to 4×10^4 kg of coal.

(b) 1 kg of uranium has an energy value of 6×10^{10} kJ and is equivalent to 2×10^6 kg of coal.

Now the energy potential of uranium can be compared with that of coal, oil and natural gas.

Table 3.5. Comparison of energy reserves
(World excluding Communist countries)

All values in billions of tonnes of coal equivalent

Coal*	1520	Proved reserves of hard coal and lignite
Oil*	140	Proved reserves
Gas*	115	Proved reserves
Uranium (a)†	146	RAR + EAR-1 at less than $130/kg U in thermal reactors
Uranium (b)†	7300	RAR + EAR-1 at less than $130/kg U in breeder reactors

Data from: *UN Energy Statistics Year Book, 1984
†Eurostat Energy Statistics Year Book, 1982

Table 3.5 shows that in existing thermal reactors, which comprise the vast majority of the world's power reactors in 1987, the contribution which uranium can make to world energy supplies is quite modest. It is comparable with the world's proved reserves of oil, and slightly greater than the world's proved reserves of natural gas. These two fossil fuels will be considerably depleted within 50 years at the present rate of consumption. Similarly, the uranium in categories RAR and EAR-1 at costs up to $130/kg U will be nearly used up within 50 years, assuming an increase in the consumption rate from the present 40 000 tonnes/year to possibly 60 000 tones/year by the year 2000 and 100 000 tonnes/year by 2010. However, by that time further reserves will no doubt become available.

On the other hand, if breeder reactors are successfully introduced on a large scale, without posing any serious operational or safety problems, then uranium has enormous potential, and will be capable of providing energy for many centuries. The large existing stockpiles of depleted uranium, i.e. the tails from enrichment plants and spent fuel from thermal reactors, will with plutonium 239 become the fuel for future breeder reactors.

For example, there was in the United Kingdom in 1980 a stockpile of about twenty thousand tonnes of depleted uranium. The energy value of this uranium when used in breeder reactors is approximately the same as that of the forty billion tonnes of coal which is believed to be ultimately recoverable in the United Kingdom. By the end of the present century this stockpile is likely to be four times larger; it is an energy resource which cannot be disregarded in a world of increasing energy consumption and decreasing resources.

The justification, therefore, for the development and use of nuclear power is that in breeder reactors uranium will be capable of providing energy for many centuries. This will not be without its dangers, for the widespread use of nuclear power and the production of plutonium on a large scale have several hazardous aspects as were demonstrated at the serious accidents at nuclear power stations at Windscale, UK in 1957, Three Mile Island, USA in 1979 and Chernobyl, USSR in 1986, and these hazards will be described in a later chapter. As was stated at the beginning of this section, the use of nuclear power on a large scale will depend on public acceptance of this source of energy, with its several hazards, and it may well be that public opinion will have the final say in the determination of the future of nuclear energy.

Chapter 4

The theory of nuclear reactors— homogeneous thermal reactors

Nuclear reactor theory is concerned with the analysis of all the processes which take place in the core of a reactor, and in particular with the slowing down, diffusion and absorption of neutrons. Only by analysing these processes in detail can accurate calculations be made to determine the multiplication factor, critical mass of fuel or size of a reactor. The complete analysis of these processes is extremely difficult and complicated, and it is necessary, particularly in a book of this nature, to simplify the theory of nuclear reactors and develop approximate models for their analysis.

In this chapter we will describe some of the models in use for thermal reactors, and apply them to the analysis of homogeneous reactors. Infinite reactors will be considered first in order to simplify the problem by neglecting neutron leakage. Subsequently neutron diffusion and slowing down will be studied and applied to finite reactors to calculate critical size and mass. It should be emphasized that calculations based on the approximate models that we will use cannot be expected to give exact answers, but they are at least of first order accuracy. More exact and elaborate reactor theories require computers for the solution of numerical problems, and a brief introduction to this topic is given in Appendix 3.

4.1 Homogeneous and quasi-homogeneous reactors

The core of a homogeneous reactor, as the name implies, is a uniform mixture of fuel, moderator and possibly other materials. For example, the core of the homogeneous aqueous reactor consists of a solution of uranyl sulphate, UO_2SO_4, in water. Reactors of this type are rather uncommon, and the vast majority of thermal reactors built up to 1987 have solid fuel elements arranged in a lattice within the moderator, which may be either solid or liquid. These reactors are clearly not homogeneous in the true sense, but if the characteristic dimensions of the fuel and its lattice (for example, the diameter of the fuel rods and the spacing between them) are smaller than the neutron mean free paths in the fuel and moderator, then the reactor is effectively homogeneous as far as neutrons are concerned. Reactors of this type are referred to as quasi-homogeneous, and can be analysed as if they are homogeneous. On the other hand, if the fuel or lattice dimensions are greater than the neutron mean free paths, the reactor must be treated as heterogeneous, and this type of reactor will be discussed in the next chapter. The theory of homogeneous reactors which is developed in this chapter can be applied to quasi-homogeneous reactors, and is therefore of more general application than might appear at first sight.

4.2 Reactor models

As we have already seen, the lifetime of a neutron in a reactor consists firstly of a period during which it is being slowed down by a number of scattering collisions. The neutron's energy is reduced at each scattering collision, and remains constant between collisions. When the neutron is thermalized it continues to diffuse in the reactor and its energy, on average, remains constant until eventually it is absorbed or leaks out of the reactor. Neutrons may also be absorbed or leak out of the reactor before they have a chance to become thermalized. The complete analysis of the slowing down and diffusion processes is very complex, and the following models have been developed to simplify matters:

1. *The one-group model.* In this model it is assumed that all neutrons are produced at thermal energy and remain at this energy throughout their lifetime, and the slowing down process is ignored. Clearly this model has serious limitations and is unlikely to give accurate solutions to reactor problems; its main advantage lies in its simplicity. It is possible, however, to make modifications to the one-group model to enable the capture and leakage of fast neutrons to be taken into account, and thereby improve the accuracy of the model.

2. *The two-group model.* In this model it is assumed that all neutrons above thermal energy can be treated as a single group, called the fast group, having a common energy. All thermal neutrons are taken as being in the second group. Suitable cross-sections for the fast group enable the rate of capture of neutrons in that group, and the rate of scattering from the fast group to the thermal group to be determined. Two-group theory represents a considerable improvement over one-group theory for reactor calculations.

3. *Multi-group models.* In these models the neutrons in the reactor are divided into a number of groups, each representing a particular energy range. In a thermal reactor one or more of these groups would include the thermal neutrons. As in two-group theory suitable cross-sections determine the rates at which neutrons are captured in each

group, and are scattered from one group to another. The mathematical complexity of these models increases as the number of groups increases, and computers are required for the solution of problems.

4. *The continuous slowing-down model.* The principal assumptions of this model are that all neutrons lose exactly ξ units of (log energy) at each scattering collision, and that the neutron density is a continuous function of energy. The first assumption implies that all neutrons behave as "average" neutrons as far as their energy loss per scattering collision is concerned. In a reactor in which neutrons are produced with a spectrum of energies by fission, and slow down by many scattering collisions in a moderator such as graphite, these assumptions are more or less correct. In a hydrogenous moderator, however, in which neutrons may be thermalized after a few collisions, the departure from these assumptions may be considerable, and the continuous slowing down model is not reliable for water moderated reactors.

4.3 The infinite homogeneous reactor

The infinite multiplication factor, k_∞, was defined in the last chapter, and we will now develop expressions for calculating the four factors of k_∞.

Eta

The factor eta was introduced in the last chapter, and its value for thermal neutron fission in natural uranium was calculated. A general expression for this factor for thermal fission in natural or enriched uranium is:

$$\eta(U) = \frac{\nu N(^{235}U)\bar{\sigma}_f(^{235}U)}{N(^{235}U)\bar{\sigma}_a(^{235}U) + N(^{238}U)\bar{\sigma}_c(^{238}U)} \quad (4.1)$$

$N(^{235}U)$ and $N(^{238}U)$ are the numbers of atoms of ^{235}U and ^{238}U, and the cross-sections are average values for thermal neutrons. Natural uranium contains 99·285 per cent of ^{238}U, and the value of $\eta(\text{nat U})$ is 1·32. The variation of $\eta(U)$ with enrichment is shown in Figure 4.1, from which it can be seen that it increases with enrichment and reaches the value 2·07 for ^{235}U.

4.4 The thermal utilization factor

The thermal utilization factor has already been defined as:

$$f = \frac{\text{The number of thermal neutrons absorbed in the fuel}}{\text{Total number of thermal neutrons absorbed in the reactor}}$$

thus:

$$f = \frac{N_F\bar{\sigma}_{aF}\phi_{thF}}{N_F\bar{\sigma}_{aF}\phi_{thF} + N_M\bar{\sigma}_{cM}\phi_{thM} + N_i\bar{\sigma}_{ci}\phi_{thi}} \quad (4.2)$$

where N_F, N_M and N_i are the numbers of atoms of fuel, moderator and

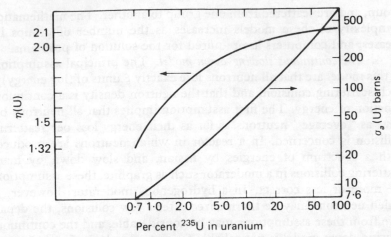

Figure 4.1. Eta and the microscopic absorption cross-section of enriched uranium

other materials per cm^3 of the reactor. In a homogeneous reactor, in which all elements in the core are completely mixed, $\phi_{thF} = \phi_{thM} = \phi_{thi}$. If it is assumed that all materials have non-$(1/v)$ factors equal to 1, the preceding expression for f becomes:

$$f = \frac{N_F \sigma_{aF}}{N_F \sigma_{aF} + N_M \sigma_{cM} + N_i \sigma_{ci}} \qquad (4.3)$$

where the cross-sections are tabulated (2200 m/s) values.

Recalling the definition of macroscopic cross-sections, $N_F \sigma_{aF}$ is Σ_{aF}, the macroscopic absorption cross-section of the fuel component of the reactor, and the denominator in equation (4.3) is the macroscopic absorption cross-section of the mixture of all the materials in the core of the reactor, which may be given the symbol Σ_{ac}. Thus the thermal utilization factor may be expressed in terms of macroscopic cross-sections as:

$$f = \frac{\Sigma_{aF}}{\Sigma_{ac}} \qquad (4.4)$$

The value of σ_{aF} in equation (4.3) is calculated by the method illustrated in Chapter 2, and its value for natural uranium is 7·6 barns. A general expression for $\sigma_a(U)$ for natural and enriched uranium is:

$$\sigma_a(U) = \frac{N(^{235}U)\sigma_a(^{235}U) + N(^{238}U)\sigma_c(^{238}U)}{N(^{235}U) + N(^{238}U)} \qquad (4.5)$$

The value of $\sigma_a(U)$ increases with enrichment as shown in Figure 4.1, reaching the value of 680 barns for pure ^{235}U. This increase in the value of $\sigma_a(U)$ has the effect of increasing the thermal utilization factor, provided the fuel-moderator ratio and the amounts of other materials remain constant.

In a water moderated reactor, or a reactor containing significant quantities of structural materials such as stainless steel which like water has rather a high capture cross-section, the use of enriched uranium is necessary to maintain a sufficiently high value of the thermal utilization factor to make criticality possible.

The variation of thermal utilization factor with moderator to fuel ratio for a natural uranium, graphite moderated reactor is shown in Figure 4.3.

According to the one-group model, in which the existence of fast neutrons and slowing down are neglected, the resonance escape probability and the fast fission factor are both equal to 1, and therefore:

$$k_\infty = f\eta \qquad (4.6)$$

This result is fairly accurate for reactors fuelled with highly enriched uranium or pure ^{235}U, in which the fast fission and resonance capture effects are negligible. It is certainly not accurate for reactors fuelled with natural or slightly enriched uranium in which the amount ^{238}U is large compared with the ^{235}U, and the fast fission and (to a greater extent) the resonance capture effects are significant. One method of calculating the resonance escape probability makes use of the continuous slowing-down model.

4.5 The continuous slowing-down model and resonance escape probability

At the outset we will define a quantity known as the neutron slowing-down density, $q(E)$, as the rate at which neutrons slow down past energy E per unit volume.

According to our analysis of elastic scattering in Chapter 2 the average decrease of (log energy) per collision, ξ, is constant. The continuous slowing-down model assumes that all neutrons behave in the average fashion and that for all neutrons the decrease of (log energy) per collision is exactly ξ at all energies. Consider an interval of (log energy), $d(\log E)$, at which the slowing-down density is $q(E)$. The fraction of neutrons which, while slowing down, is scattered into the interval $d(\log E)$ is $d(\log E)/\xi$, which is equal to $dE/\xi E$.

The total number of neutrons scattered into the interval dE at E is therefore:

$$\frac{q(E)\,dE}{\xi E} \quad \text{neutrons/cm}^3\text{s}$$

In a purely scattering medium, in which $\sigma_c = 0$, the number of neutrons scattered into dE per second is equal to the number of neutrons scattered out of dE per second. If $\phi(E)dE$ is the flux of neutrons whose energy is in the range E to $E + dE$, the rate of scattering out of dE is $\Sigma_s\phi(E)\,dE$.

We now have:

$$\frac{q(E)\,\mathrm{d}E}{\xi E} = \Sigma_s \phi(E)\,\mathrm{d}E$$

or
$$\phi(E) = \frac{q(E)}{\xi \Sigma_s E} \tag{4.7}$$

In a pure scattering medium in which no neutrons are captured, the slowing-down density in a steady-state situation is constant at all energies. If Σ_s is assumed to be constant, and noting that ξ is independent of energy, it follows that the neutron flux per unit energy at energy E is inversely proportional to E:

$$\phi(E) \propto \frac{1}{E} \tag{4.8}$$

This result is only true for an infinite scattering medium, however in a thermal reactor the scattering cross-section of the fuel-moderator mixture is much greater than the capture cross-section. (If this were not so, neutrons would not be thermalized and a chain reaction would be impossible.) Equation (4.8) represents quite accurately, therefore, the variation of neutron flux with energy in a thermal reactor down to the upper end of the thermal neutron spectrum.

In a scattering and absorbing medium the number of neutrons scattered into the energy interval $\mathrm{d}E$ at E is still $[q(E)\,\mathrm{d}E]/\xi E$, although the slowing-down density is no longer constant at all energies, but decreases as the neutron energy decreases due to capture during slowing down. The rate at which neutrons are scattered into the energy interval $\mathrm{d}E$ is equal to the rate at which neutrons are captured in this interval plus the rate at which they are scattered out to lower energies:

$$\frac{q(E)\,\mathrm{d}E}{\xi E} = (\Sigma_s + \Sigma_a(E))\,\phi(E)\,\mathrm{d}E \tag{4.9}$$

The rate at which neutrons are captured in $\mathrm{d}E$ is a measure of the decrease in the slowing-down density, $\mathrm{d}q(E)$:

$$\mathrm{d}q(E) = \Sigma_a(E)\phi(E)\,\mathrm{d}E \tag{4.10}$$

(Note that Σ_a is dependent on energy while Σ_s is assumed to be constant, and that $\bar{\xi}$ refers to the mixture of fuel and moderator in a reactor.)

Dividing equation (4.10) by (4.9), and rearranging:

$$\frac{\mathrm{d}q(E)}{q(E)} = \frac{1}{\bar{\xi}}\frac{\Sigma_a(E)}{\Sigma_s + \Sigma_a(E)}\frac{\mathrm{d}E}{E} \tag{4.11}$$

Integrating:

$$\log_e \frac{q(E_2)}{q(E_1)} = -\frac{1}{\bar{\xi}} \int_{E_2}^{E_1} \frac{\Sigma_a(E)}{\Sigma_s + \Sigma_a(E)} \frac{dE}{E}$$

or

$$\frac{q(E_2)}{q(E_1)} = \exp\left[-\frac{1}{\bar{\xi}} \int_{E_2}^{E_1} \frac{\Sigma_a(E)}{\Sigma_s + \Sigma_a(E)} \frac{dE}{E} \right] \quad (4.12)$$

If E_1 and E_2 are the upper and lower limits of the resonance range, then $q(E_2)/q(E_1)$ is the fraction of neutrons which, while slowing down, escape capture in the resonance absorber, that is the resonance escape probability. To include neutron capture at all energies from fission to thermal, E_1 should be taken as fission energy and E_2 as thermal energy. Now:

$$p = \exp\left[-\frac{1}{\bar{\xi}} \int_{E_{th}}^{E_f} \frac{\Sigma_a(E)}{\Sigma_s + \Sigma_a(E)} \frac{dE}{E} \right] \quad (4.13)$$

This expression for the resonance escape probability is the starting point for several different methods of calculation. The problem in using it lies in evaluating the integral, particularly in view of the fact that the capture cross-section of ^{238}U varies widely in the resonance range, and it is quite impossible to represent this variation by any relationship between Σ_a and E.

One method of evaluating the integral in equation (4.13) which relies on experimental measurements of the cross-sections of ^{238}U (or ^{232}Th, the other possible resonance absorber) will be outlined. If it is assumed that in the fuel-moderator mixture ^{238}U is the only significant absorber of resonance neutrons, then $\Sigma_a(E) = N(^{238}U)\sigma_c(^{238}U)$ where $N(^{238}U)$ is the number of atoms of ^{238}U/cm^3 of the mixture. Equation (4.13) can be written as:

$$p = \exp\left[-\frac{N(^{238}U)}{\bar{\xi}\Sigma_s} \int_{E_{th}}^{E_f} \frac{\sigma_c(^{238}U)}{1 + \dfrac{N(^{238}U)\sigma_c(^{238}U)}{\Sigma_s}} \frac{dE}{E} \right] \quad (4.14)$$

The only variable in this equation for the resonance escape probability is $N(^{238}U)/\Sigma_s$, or its reciprocal $\Sigma_s/N(^{238}U)$ which is the total scattering cross-section of the fuel-moderator mixture per atom of ^{238}U, regardless of the identity of the moderator. The integral of equation (4.14) is known as the effective resonance integral, I:

$$I = \int_{E_{th}}^{E_f} \frac{\sigma_c(^{238}U)}{1 + \dfrac{N(^{238}U)}{\Sigma_s} \sigma_c(^{238}U)} \frac{dE}{E} \quad (4.15)$$

For pure ^{238}U:

$$I = \int_{E_{th}}^{E_f} \frac{\sigma_c(^{238}U)}{1 + \dfrac{\sigma_c(^{238}U)}{\sigma_s(^{238}U)}} \frac{dE}{E} \qquad (4.16)$$

which has been experimentally determined to be about 9 barns, and $\Sigma_s/N(^{238}U) = \sigma_s(^{238}U)$ which is 8·3 barns.

For very large values of the moderator to fuel ratio, $\Sigma_s/N(^{238}U)$ becomes very large, and in the limit as $\Sigma_s/N(^{238}U) \to \infty$, the effective resonance integral becomes:

$$I = \int_{E_{th}}^{E_f} \sigma_c(^{238}U) \frac{dE}{E} \qquad (4.17)$$

which has been experimentally determined to be about 280 barns.

The available data for the variation of I with $\Sigma_s/N(^{238}U)$ for values of $\Sigma_s/N(^{238}U)$ up to about 10 000 barns is best represented by the empirical equation:

$$I = 2 \cdot 73 \left\{ \frac{\Sigma_s}{N(^{238}U)} \right\}^{0.486} \text{ barns} \qquad (4.18)$$

This variation is shown in Figure 4.2.

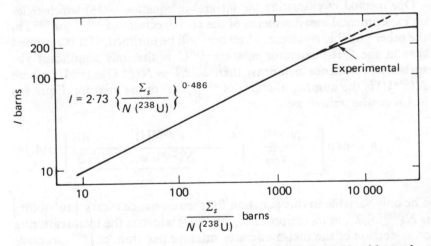

Figure 4.2. The variation of effective resonance integral with total scattering cross-section for ^{238}U

Inserting the expression for the effective resonance integral, equation (4.18), into the expression for the resonance escape probability, equation (4.14), we have:

$$p = \exp\left[-\frac{2 \cdot 73}{\bar{\xi}}\left\{\frac{\Sigma_s}{N(^{238}\text{U})}\right\}^{-0 \cdot 514}\right] \qquad (4.19)$$

This equation can be used for computational purposes for homogeneous systems of natural or slightly enriched uranium with any moderator. The value of $\bar{\xi}$ should be an average value for the fuel–moderator mixture, but the difference between this average value and the value for the pure moderator is likely to be very small and is usually negligible.

The variation of resonance escape probability with moderator to fuel ratio for a homogeneous mixture of natural uranium and graphite is shown in Figure 4.3.

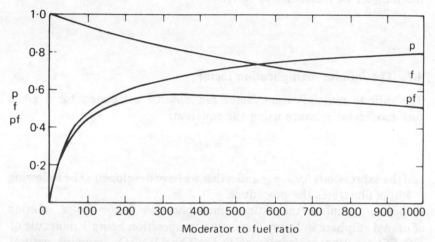

Figure 4.3. Variation of resonance escape probability and thermal utilization factor of a natural uranium and graphite mixture

4.6 The fast fission factor

For dilute mixtures of fuel and moderator (in which the ratio of moderator atoms or molecules to fuel atoms is greater than, say, 50 to 1), the fast fission factor, ϵ, may be taken as equal to 1 without the need for detailed calculation. The reason for this is that when fission neutrons are emitted at an average energy of 2 MeV in a medium containing a small quantity òf fuel and a large quantity of moderator the most probable type of interaction is elastic scattering in the moderator. There may be a little inelastic scattering in the fuel, but negligibly few neutrons will cause fission in the fuel. As a result the neutrons' energy is reduced below 1 MeV, the fission threshold in ^{238}U, without any appreciable increase in the number of neutrons due to fast fission.

In the case of certain water moderated reactors which use enriched uranium as the fuel, the moderator–fuel ratio may be less than the figure quoted above, and the fast fission factor may be significant. Calculations for this type of reactor have given the following formula for ϵ:

$$\epsilon = \frac{1 + 0.690 \left\{ \dfrac{N(^{238}U)}{N(H_2O)} \right\}}{1 + 0.563 \left\{ \dfrac{N(^{238}U)}{N(H_2O)} \right\}} \tag{4.20}$$

where $N(^{238}U)/N(H_2O)$ is the ratio of the number of atoms of ^{238}U to the number of molecules of water.

4.7 The infinite multiplication factor

The infinite multiplication factor can now be calculated for a given fuel–moderator mixture using the equation:

$$k_\infty = \epsilon p f \eta$$

and the expressions for ϵ, p, f and η that we have developed. The following example illustrates the procedure.

Example: Calculate the infinite multiplication factor for a solution of uranyl sulphate in heavy water, the composition being 1 molecule of UO_2SO_4 to 1000 molecules of D_2O. The UO_2SO_4 contains natural uranium.

In this example we can base our calculation on any quantity of fuel–moderator mixture that is convenient, and in this case the simplest quantity is one molecule of fuel and 1000 molecules of moderator.

The average value of the logarithmic energy decrement for D_2O is found using equation (2.25):

$$\begin{aligned}
\bar{\xi}(D_2O) &= \frac{N(D)\sigma_s(D)\xi(D) + N(O)\sigma_s(O)\xi(O)}{N(D)\sigma_s(D) + N(O)\sigma_s(O)} \\
&= \frac{2 \times 3.4 \times 0.725 + 1 \times 3.8 \times 0.120}{2 \times 3.4 + 1 \times 3.8} \\
&= 0.509
\end{aligned}$$

The calculation may be carried out in tabular form as follows:

Sub-stance	At. Wt.	N	σ_a	σ_s	$N\sigma_a$	$N\sigma_s$	ξ	$N\sigma_s\xi$
U	238	1	7·6	8·3	7·56	8·3	0·0084	0·0697
O	16	6	0	3·8	0	22·8	0·1209	2·755
S	32	1	0·52	1·1	0·52	1·1	0·0616	0·0678
D$_2$O	—	1000	0·001	10·6	1·00	10 600	0·509	5395·4
					9·08	10 632		5398·3
					$\sum N\sigma_a$	$\sum N\sigma_s$		$\sum N\sigma_s\xi$

From equation (4.3):

$$f = \frac{N(\text{U})\sigma_a(\text{U})}{\sum N\sigma_a} = \frac{7\cdot56}{9\cdot08} = 0\cdot833$$

From equation (2.25):

$$\bar{\xi} = \frac{\sum N_i\sigma_{si}\xi_i}{\sum N_i\sigma_{si}} = \frac{5398\cdot3}{10\ 632} = 0\cdot508$$

This result illustrates the point mentioned earlier that the value of $\bar{\xi}$ for a dilute fuel–moderator mixture is very nearly the same as for pure moderator.

From the values in the table above:

$$\frac{\Sigma_s}{N(^{238}\text{U})} = \frac{10\ 632}{0\cdot992\ 85} = 10\ 710 \text{ barns per atom of } ^{238}\text{U}$$

From equation (4.19):

$$p = \exp\left[-\frac{2\cdot73 \times 10\ 710^{-0\cdot514}}{0\cdot508}\right]$$

$$= \exp(-0\cdot046\ 25) = 0\cdot955$$

Combining these results:

$$k_\infty = 1 \times 0\cdot955 \times 0\cdot833 \times 1\cdot32$$

$$= 1\cdot05$$

This result indicates that a chain reaction is possible with an infinite system of natural uranium (as UO_2SO_4) in heavy water for the moderator to fuel ratio chosen. In the case of graphite or light water as moderator, similar calculations show that in an infinite homogeneous system with natural uranium as fuel, criticality is impossible for any moderator to fuel ratio.

For a given fuel, an increase in the moderator to fuel ratio has the effect of decreasing f and increasing p, while leaving the values of ϵ and η unchanged. This situation is shown in Figure 4.3 for a natural uranium and graphite mixture. The variation of the product pf is also shown and it can be seen that this quantity, and hence k_∞, has a maximum value corresponding to an optimum moderator to fuel ratio. Figure 4.3 also illustrates the point just mentioned that a chain reaction in a natural uranium–graphite system is impossible, even at the optimum moderator to fuel ratio. The maximum value of the product pf for this system is (referring to Figure 4.3) just less than 0·6, and the maximum value of k_∞ is less than 0·75.

4.8 The effect of enrichment

It is necessary in the case of homogeneous water and graphite moderated reactors to use enriched uranium to achieve criticality. The effect of enrichment on the infinite multiplication factor can best be demonstrated by calculation, and the example of a graphite moderated reactor in which the moderator to fuel ratio is 300 will be considered. As shown in Figure 4.3, this moderator to fuel ratio gives the optimum value of the infinite multiplication factor when the fuel is natural uranium. It will be assumed for the purposes of this example that the value of the fast fission factor is one and is not affected by enrichment.

Table 4.1. The infinite multiplication factor for a
homogeneous uranium–graphite reactor

Fraction of ^{235}U in the fuel	σ_a (U)	f	η (U)	p	k_∞
0·007 15 (Nat U)	7·56	0·848	1·325	0·659	0·741
0·01	9·49	0·875	1·476	0·659	0·852
0·015	12·88	0·905	1·632	0·659	0·973
0·02	16·27	0·923	1·722	0·659	1·048
0·03	23·04	0·945	1·824	0·659	1·135

Note that the effect of enrichment is to increase the values of σ_a(U), η(U) and f while leaving p virtually unchanged. The overall effect is to increase the infinite multiplication factor, and make possible criticality in a uranium–graphite reactor with slightly enriched uranium containing approximately 2 per cent ^{235}U. A similar result can be demonstrated for a water moderated uranium fuelled reactor.

4.9 Finite homogeneous reactors

The restriction of our discussion so far to infinite reactors has been in the interests of simplicity, to avoid having to consider neutron diffusion and leakage. It goes without saying that no such thing as an infinite reactor exists, and all reactor cores are of finite size, varying from a small assembly of pure ^{235}U less than the size of a football to very large cores of dimensions 10 metres or more.

4.10 One-group neutron diffusion theory

The calculation of neutron leakage from a reactor core depends on a knowledge of the way in which neutrons move about or diffuse in the reactor both during slowing down and while they are thermal. The simplest theory of neutron diffusion is the one-group theory, mentioned earlier in this chapter, in which it is assumed that all neutrons in the reactor are at thermal energy. The effect of scattering collisions on these neutrons is to alter their direction of motion, but not their speed.

In connection with neutron motion we define a quantity called the neutron current density, $\mathcal{J}$, as follows:

In a Cartesian coordinate system (see Figure 4.4) the partial neutron current density in the positive z-direction, $\mathcal{J}_{z+}$, is the number of neutrons per second which pass upwards through a unit area parallel to the x–y plane. The partial neutron current density in the negative z-direction, $\mathcal{J}_{z-}$, is the number of neutrons per second which pass downwards through a unit area parallel to the x–y plane. The neutron current density in the z-direction, $\mathcal{J}_z$, is given by:

$$\mathcal{J}_z = \mathcal{J}_{z+} - \mathcal{J}_{z-} \qquad (4.21)$$

In order to derive an expression for the neutron current density in terms of the neutron flux we will make the following assumptions:

1. The medium in which neutrons are diffusing is homogeneous and isotropic, and neutron scattering is isotropic in the L system. This implies that neutrons scattered at any point emerge isotropically from that point. A correction can be made to the result of this analysis to take account of non-isotropic scattering in the L system.

2. The scattering cross-section of the medium is much larger than the absorption cross-section.

3. The neutron flux is a slowly varying function of position. This will be the case if $\Sigma_a \ll \Sigma_s$, and at points a few mean free paths away from the boundary of the medium, and neutron sources.

One-group diffusion theory, which is based on these assumptions, provides a simple and fairly accurate method for analysing neutron diffusion, and in practice it is not necessary to impose too strictly the restrictions implied by these assumptions. Consider an infinite medium in which neutrons are diffusing and being scattered, with an element of volume dV whose position is defined by the vector $\mathbf{r}$, and an element of area dA lying in the x–y plane at the origin of the coordinate system, Figure 4.4.

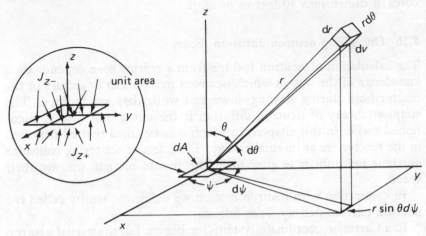

Figure 4.4. Neutron scattering and current

If the neutron flux at position $\mathbf{r}$ is $\phi(\mathbf{r})$, the rate of scattering in dV per second is $\Sigma_s \phi(\mathbf{r})\, \mathrm{d}V$.

The fraction of neutrons which are scattered in dV into the direction towards dA is $(\mathrm{d}A \cos \theta)/4\pi r^2$, r being the distance from dV to dA.

The fraction of neutrons which, having been scattered in dV towards dA, reach and pass through dA without further interaction is $e^{-\Sigma_t r}$. Since the validity of diffusion theory depends, as stated above, on Σ_a being much less than Σ_s, this term can be written as $e^{-\Sigma_s r}$.

Combining these three quantities, the number of neutrons which are scattered in dV and then pass through dA per second is:

$$\frac{\Sigma_s \phi(\mathbf{r})\, e^{-\Sigma_s r} \cos \theta\, \mathrm{d}V\, \mathrm{d}A}{4\pi r^2}$$

The element of volume dV may be expressed in spherical coordinates as:

$$\mathrm{d}V = r^2 \sin \theta\, \mathrm{d}r\, \mathrm{d}\theta\, \mathrm{d}\psi$$

The total number of neutrons which pass through dA per second from above to below, that is in the negative z-direction, is obtained by integrating the expression:

$$\frac{\Sigma_s\, \mathrm{d}A}{4\pi}\, \phi(\mathbf{r})\, e^{-\Sigma_s r} \cos \theta \sin \theta\, \mathrm{d}r\, \mathrm{d}\theta\, \mathrm{d}\psi$$

over the whole of space above the x–y plane. In fact only the region within about four or five mean free paths of the origin contributes to the integral because of the term $e^{-\Sigma_s r}$. The expression we derive for neutron current by considering an infinite medium can therefore be applied to a finite medium at all points more than a few mean free paths from its boundary. This condition was noted in assumption 3 above.

To evaluate the integral of the preceding expression, $\phi(\mathbf{r})$ must be expressed in terms of ϕ_0, the flux at the origin where the neutron current is being evaluated. Expressing $\phi(\mathbf{r})$ in terms of ϕ_0 and its derivatives by means of a Taylor series, and bearing in mind that ϕ is assumed to be a slowly varying function, we have:

$$\phi(\mathbf{r}) = \phi_0 + x\left(\frac{\partial \phi}{\partial x}\right)_0 + y\left(\frac{\partial \phi}{\partial y}\right)_0 + z\left(\frac{\partial \phi}{\partial z}\right)_0 + \cdots$$

$$= \phi_0 + r \sin \theta \cos \psi \left(\frac{\partial \phi}{\partial x}\right)_0 + r \sin \theta \sin \psi \left(\frac{\partial \phi}{\partial y}\right)_0 + r \cos \theta \left(\frac{\partial \phi}{\partial z}\right)_0$$

When this expression for $\phi(\mathbf{r})$ is inserted in the integral the terms involving ψ become zero as the integration with respect to ψ is carried out from 0 to 2π. The partial neutron current in the negative z-direction can be determined by dividing the total number of neutrons passing through $\mathrm{d}A$ from above by $\mathrm{d}A$, giving:

$$\mathcal{J}_{z-} = \frac{\Sigma_s}{4\pi} \int_{\theta=0}^{\pi/2} \int_{\psi=0}^{2\pi} \int_{r=0}^{\infty} \left\{ \phi_0 + r \cos \theta \left(\frac{\partial \phi}{\partial z}\right)_0 \right\} e^{-\Sigma_s r} \cos \theta \sin \theta \, \mathrm{d}r \, \mathrm{d}\psi \, \mathrm{d}\theta$$

The solution of this integral gives:

$$\mathcal{J}_{z-} = \frac{\phi_0}{4} + \frac{1}{6\Sigma_s}\left(\frac{\partial \phi}{\partial z}\right)_0 \qquad (4.22)$$

The expression for $\mathcal{J}_{z+}$ is obtained by the same method, with θ varying from $\pi/2$ to π, and is:

$$\mathcal{J}_{z+} = \frac{\phi_0}{4} - \frac{1}{6\Sigma_s}\left(\frac{\partial \phi}{\partial z}\right)_0 \qquad (4.23)$$

The neutron current density in the z-direction is obtained by subtracting equation (4.22) from (4.23)

$$\mathcal{J}_z = \mathcal{J}_{z+} - \mathcal{J}_{z-} = -\frac{1}{3\Sigma_s}\left(\frac{\partial \phi}{\partial z}\right)_0 \qquad (4.24)$$

The neutron current density in the x and y directions are given by similar expressions, namely:

$$\mathcal{J}_x = -\frac{1}{3\Sigma_s}\left(\frac{\partial\phi}{\partial x}\right) \quad \text{and} \quad \mathcal{J}_v = -\frac{1}{3\Sigma_s}\left(\frac{\partial\phi}{\partial y}\right)$$

In general:

$$\mathcal{J} = -\frac{1}{3\Sigma_s}\,\text{grad}\,\phi \tag{4.25}$$

where:

$$\text{grad}\,\phi = \mathbf{i}\frac{\partial\phi}{\partial x} + \mathbf{j}\frac{\partial\phi}{\partial y} + \mathbf{k}\frac{\partial\phi}{\partial z}$$

The neutron current is proportional to the gradient of the flux, and the flow of neutrons is in the direction of decreasing flux. The constant of proportionality in equation (4.25) is called the diffusion coefficient, D, and this equation may be written as:

$$\mathcal{J} = -D\,\text{grad}\,\phi \tag{4.26}$$

where:

$$D = \frac{1}{3\Sigma_s} = \frac{\lambda_s}{3} \tag{4.27}$$

One of the assumptions upon which these results are based is that of isotropic scattering in the L system which we know to be incorrect, especially for scattering with light elements such as hydrogen. If scattering is isotropic in the C system, but anisotropic in the L system, it is possible by the methods of transport theory to show (in the case where Σ_s is much larger than Σ_a) that the diffusion coefficient is given by:

$$D = \frac{1}{3(1-\bar{\mu})\Sigma_s} = \frac{1}{3\Sigma_{tr}} = \frac{\lambda_{tr}}{3} \tag{4.28}$$

Σ_{tr} and λ_{tr} are called the transport cross-section and the transport mean free path respectively, and:

$$\Sigma_{tr} = \Sigma_s(1-\bar{\mu}) \tag{4.29}$$

where $\bar{\mu}$ is the average cosine of the scattering angle in the L system (refer to Chapter 2). Equation (4.28) is preferred to (4.27) for expressing the diffusion coefficient in terms of the scattering cross-section.

We can now use the equation for neutron current density to determine the neutron leakage out of unit volume in a diffusing medium. Considering an element of volume in a Cartesian coordinate system (see Figure 4.5) we can use the result of the preceding section as follows:

The flow of neutrons into the element of volume through the lower

face parallel to the x–y plane $= -D(\partial\phi/\partial z)_z\ dx\ dy$.
The flow of neutrons out of the element of volume through the upper face parallel to the x–y plane $= -D(\partial\phi/\partial z)_{z+dz}\ dx\ dy$.

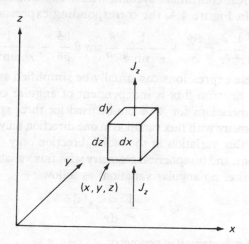

Figure 4.5. Neutron leakage from an element of volume

$(\partial\phi/\partial z)$ at coordinate $z + dz$ can be expressed in terms of $(\partial\phi/\partial z)$ at z by the first two terms of a Taylor series as:

$$\left(\frac{\partial\phi}{\partial z}\right)_{z+dz} = \left(\frac{\partial\phi}{\partial z}\right)_z + \left(\frac{\partial^2\phi}{\partial z^2}\right)_z dz$$

Using this equation, the net flow of neutrons out of the element of volume through the faces parallel to the x–y plane is:

$$-D\left(\frac{\partial^2\phi}{\partial z^2}\right)_z dx\ dy\ dz$$

Noting that $dx\ dy\ dz$ is the volume of the element, and that similar expressions will give the rate of leakage through the other faces of the element of volume, the total leakage of neutrons per second out of a unit volume in a Cartesian coordinate system is:

$$\text{Leakage} = -D\left(\frac{\partial^2\phi}{\partial x^2} + \frac{\partial^2\phi}{\partial y^2} + \frac{\partial^2\phi}{\partial z^2}\right)$$

$$= -D\ \nabla^2\phi \tag{4.30}$$

where ∇^2 is the Laplacian operator.

Equation (4.30) is a general equation for the rate of neutron leakage from unit volume in any coordinate system. In a cylindrical coordinate system the corresponding expression for $\nabla^2\phi$ is:

$$\nabla^2\phi = \left(\frac{1}{r}\frac{\partial}{\partial r}r\frac{\partial\phi}{\partial r} + \frac{1}{r^2}\frac{\partial^2\phi}{\partial\theta^2} + \frac{\partial^2\phi}{\partial z^2}\right) \tag{4.31}$$

and in a spherical coordinate system, whose coordinates are the same as those used in Figure 4.4, the corresponding expression for $\nabla^2\phi$ is:

$$\nabla^2\phi = \left(\frac{1}{r^2}\frac{\partial}{\partial r}r^2\frac{\partial\phi}{\partial r} + \frac{1}{r^2\sin\theta}\frac{\partial}{\partial\theta}\sin\theta\frac{\partial\phi}{\partial\theta} + \frac{1}{r^2\sin^2\theta}\frac{\partial^2\phi}{\partial\psi^2}\right) \tag{4.32}$$

In practice these expressions can usually be simplified as in most cases of interest the neutron flux is independent of angular coordinate.

Simplified expressions for $\nabla^2\phi$ can be used for three special cases, (a) rectangular geometry with flux variation in one direction only, (b) cylindrical geometry with flux variation in the radial direction only (i.e. no angular or axial variation), and (c) spherical geometry with flux variation in the radial direction only (i.e. no angular variation) as follows:

$$\nabla^2\phi = \frac{1}{r^n}\frac{\mathrm{d}}{\mathrm{d}r}r^n\frac{\mathrm{d}\phi}{\mathrm{d}r}$$

where $n = 0$ for rectangular geometry,
$\quad\quad n = 1$ for cylindrical geometry,
and $\quad n = 2$ for spherical geometry.

4.11   The neutron diffusion equation

We can now develop a neutron balance equation for unit volume of a medium in which neutrons are being produced, absorbed and are diffusing at constant energy. The rate of change of the neutron density is equal to the rate at which neutrons are produced per unit volume in the medium minus the sum of the rates of neutron leakage and absorption per unit volume in the medium. The neutron diffusion equation is:

$$\frac{\partial n}{\partial t} = S - \Sigma_a\phi - (-D\nabla^2\phi) \tag{4.33}$$

where S is the source of neutrons per unit volume.

In a reactor operating at steady-state the neutron density is independent of time and equation (4.33) becomes:

$$D\nabla^2\phi - \Sigma_a\phi + S = 0 \tag{4.34}$$

This is the steady-state diffusion equation. It is the starting point for the solution of neutron diffusion problems both in multiplying media such as reactors, and in non-multiplying media in which neutrons are not being produced by fission.

The solution of the diffusion equation must satisfy certain boundary and other conditions. These are summarized below.

1. The neutron flux must be finite and non-negative at all points where the diffusion equation applies. The condition of finite flux does not necessarily apply at points where localized neutron sources exist as the diffusion equation itself is not valid at such points.
2. In a system which has a plane, line or point of symmetry, the neutron flux is symmetrical about such a plane, line or point.
3. At an interface between two different media the neutron current density normal to the interface and the neutron flux are both continuous across the interface.
4. At the free surface of a medium the neutron flux varies in such a way that if it is extrapolated beyond the free surface it becomes zero at a fixed distance, known as the extrapolation distance.

Conditions 1 and 2 are more or less self-evident.

If the interface between two media, A and B, referred to in condition 3 is parallel to the y–z plane at $x = 0$, then:

$$\mathcal{J}_{x+} \text{ in A} = \mathcal{J}_{x+} \text{ in B} \quad \text{at } x = 0$$

and
$$\mathcal{J}_{x-} \text{ in A} = \mathcal{J}_{x-} \text{ in B} \quad \text{at } x = 0$$

(If these equations were not correct, then there would be an accumulation or loss of neutrons at the interface which is physically impossible.)

The preceding equations may be written, referring to equations (4.22) and (4.23), as:

$$\left(\frac{\phi_A}{4}\right)_0 - \frac{1}{6\Sigma_{sA}}\left(\frac{\partial \phi_A}{\partial x}\right)_0 = \left(\frac{\phi_B}{4}\right)_0 - \frac{1}{6\Sigma_{sB}}\left(\frac{\partial \phi_B}{\partial x}\right)_0$$

and
$$\left(\frac{\phi_A}{4}\right)_0 + \frac{1}{6\Sigma_{sA}}\left(\frac{\partial \phi_A}{\partial x}\right)_0 = \left(\frac{\phi_B}{4}\right)_0 + \frac{1}{6\Sigma_{sB}}\left(\frac{\partial \phi_B}{\partial x}\right)_0$$

Adding these two equations we get:

$$\phi_A = \phi_B \quad \text{at } x = 0$$

and subtracting the first equation from the second we get:

$$\mathcal{J}_{xA} = \mathcal{J}_{xB} \quad \text{at } x = 0$$

The conditions of continuity of neutron flux and current at the interface between A and B are thus demonstrated.

If a neutron diffusing medium is bounded by a plane surface beyond which is a space of very low density, such as air (which from the point of view of neutron scattering is for all practical purposes a vacuum), then neutrons escaping from the medium through its surface into the surrounding space are not scattered and none return across the surface. The neutron current at this free surface is therefore entirely in the outward direction, or (referring to Figure 4.6 in which the free surface is normal to the x-axis) the positive x direction. Recalling equation (4.24), this implies that the flux gradient at the free surface is negative.

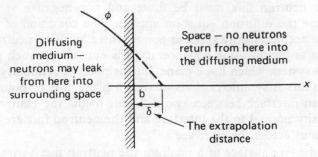

Figure 4.6. The free surface boundary condition

The neutron flux in the diffusing medium at the free surface can be assumed to vary in such a way as to satisfy the equation:

$$\frac{1}{\phi_b}\left(\frac{\partial \phi}{\partial x}\right)_b = -\frac{1}{\delta} \tag{4.35}$$

where b is the coordinate of the free surface, and δ is the distance beyond the free surface at which the flux, if extrapolated linearly beyond the free surface, becomes zero. δ is called the extrapolation distance, and b + δ the extrapolated boundary.

Referring to Figure 4.6, at the boundary of the diffusing medium, where $x = b$, the partial neutron current density in the negative x direction, $\mathcal{J}_{x-}$, is zero, and the net neutron current density $\mathcal{J}_x$ is equal to the partial neutron current density in the positive x direction, $\mathcal{J}_{x+}$. This condition may be expressed, using equations (4.23) and (4.24) as:

$$-\frac{1}{3\Sigma_s}\left(\frac{\partial \phi}{\partial x}\right)_b = \frac{\phi_b}{4} - \frac{1}{6\Sigma_s}\left(\frac{\partial \phi}{\partial x}\right)_b$$

which, using equation (4.27), gives:

$$\frac{1}{\phi_b}\left(\frac{\partial \phi}{\partial x}\right)_b = -\frac{1}{2D} \tag{4.36}$$

Comparing equations (4.35) and (4.36), it can be seen that the assumption implied in equation (4.35) is justified by diffusion theory, and the extrapolation distance δ is related to the diffusion coefficient D by the equation:

$$\delta = 2D \tag{4.37}$$

According to transport theory it can be shown that better agreement between equation (4.35) and the actual flux distribution at the free surface is given by:

$$\delta = 2 \cdot 13 \, D \tag{4.38}$$

A simpler form of boundary condition 4 which mathematically is more convenient to use, although it is not physically correct, is:

The neutron flux becomes zero at the extrapolated boundary of the medium, i.e. $\phi = 0$ at $x = b + \delta$.

For most media the diffusion coefficient is about 1 cm, and the extrapolation distance is about 2 cm. Many reactors, particularly power reactors, are a few metres in size, and the extrapolation distance can be neglected by comparison with this size. In such cases the flux may be assumed to be zero at the actual surface of the reactor.

4.12 The diffusion length

Returning to the diffusion equation, (4.34), we will consider the case of a diffusing medium in which there are localized neutron sources only. In such a medium, at all points where there are no sources, the steady-state diffusion equation is written as:

$$D \, \nabla^2 \phi - \Sigma_a \phi = 0$$

or

$$\nabla^2 \phi - \frac{1}{L^2} \phi = 0 \qquad (4.39)$$

where $L = \sqrt{D/\Sigma_a}$, and is called the diffusion length.

The diffusion length is a very important parameter in nuclear engineering and it is useful at this stage to obtain some idea of its significance. Let us consider a point source emitting S neutrons per second isotropically in an infinite medium. At all points, except in the vicinity of the neutron source, the flux is found from the solution of equation (4.39), which in terms of spherical coordinates with angular symmetry may be written as:

$$\frac{d^2 \phi}{dr^2} + \frac{2}{r} \frac{d\phi}{dr} - \frac{1}{L^2} \phi = 0 \qquad (4.40)$$

Making the substitution $x = \phi r$, this equation becomes:

$$\frac{d^2 x}{dr^2} - \frac{1}{L^2} x = 0$$

The solution of this equation is:

$$x = A \, e^{-r/L} + C \, e^{r/L}$$

or

$$\phi = \frac{A}{r} e^{-r/L} + \frac{C}{r} e^{r/L} \qquad (4.41)$$

As $r \to \infty$ the second term of this solution becomes infinite, and therefore C must be zero to satisfy boundary condition 1. The equation for the flux is, therefore:

$$\phi(r) = \frac{A}{r} e^{-r/L} \qquad (4.42)$$

The value of the constant A may be found in terms of the source strength, S, by noting that at steady state the rate of production of neutrons is equal to the total rate of absorption of neutrons in the entire infinite medium. This condition can be expressed by the equation:

$$S = \int_0^\infty \Sigma_a \phi(r) 4\pi r^2 \, dr$$

or

$$S = 4\pi \Sigma_a A \int_0^\infty e^{-r/L} \, r \, dr$$

The value of the integral is L^2, and consequently:

$$A = \frac{S}{4\pi D} \qquad (4.43)$$

and the flux is given by:

$$\phi(r) = \frac{S}{4\pi D r} e^{-r/L} \qquad (4.44)$$

The mean square crow-flight distance, $\overline{r^2}$, travelled by neutrons from their source to their point of absorption may be expressed by the equation:

$$\overline{r^2} = \frac{1}{S} \sum_{\substack{\text{all} \\ \text{neutrons}}} \left\{ \begin{array}{l} \text{The number of neutrons absorbed per second} \\ \text{in the distance } r \text{ to } r + dr \text{ from the source} \end{array} \right\} \times r^2$$

$$= \frac{1}{S} \int_0^\infty \Sigma_a \phi(r) 4\pi r^4 \, dr$$

$$= \frac{1}{L^2} \int_0^\infty e^{-r/L} \, r^3 \, dr$$

The integral is equal to $6L^4$, consequently:

$$\overline{r^2} = 6L^2$$

or

$$L^2 = \tfrac{1}{6}\overline{r^2} \qquad (4.45)$$

The diffusion length squared is one sixth of the mean square crow-flight distance travelled by neutrons which are diffusing with constant energy from their source up to their point of absorption.

The significance of L^2 as being proportional to the mean square crowflight distance travelled by neutrons in a material is important, as will be seen in a later section. At this stage it is sufficient to emphasize that the larger the value of L^2, the further neutrons travel (on average) from their source to the point where they are absorbed.

For thermal neutrons, L^2 is one-sixth or the mean square crow-flight distance travelled from the point at which neutrons become thermalized at the end of the slowing down process, to the point where they are absorbed.

4.13 Neutron diffusion in multiplying media – the reactor equation

The most important use of the one-group neutron diffusion equation is in the solution of problems associated with thermal reactors, in particular the determination of the critical size of the core and the spatial variation of the neutron flux in the core. One-group theory for thermal reactors assumes that all neutrons in the core are thermal neutrons, and the monoenergetic diffusion equation, (4.33), can be used for the analysis of thermal reactors by integrating this equation over the thermal neutron spectrum, and using appropriate values of the diffusion coefficient and the absorption cross-section.

Integrating equation (4.33) over the thermal neutron energy range we get the following equation:

$$\int_{th} \frac{\partial n(\mathbf{r}, E, t)}{\partial t} \, dE = \int_{th} D_c(E) \, \nabla^2 \phi(\mathbf{r}, E, t) \, dE$$

$$- \int_{th} \Sigma_{ac}(E) \phi(\mathbf{r}, E, t) \, dE \;\; + \int_{th} S(\mathbf{r}, E, t) \, dE \qquad (4.46)$$

If it is assumed that the neutron flux and density can be considered as separable functions of position, energy and time, for example if:

$$\phi(\mathbf{r}, E, t) = F(\mathbf{r}) G(E) T(t)$$

then the integration of equation (4.46) over thermal energies can be carried out using the following results from Chapter 2, in which the variable v has been replaced by E without affecting the results:

$$\int_{th} \phi(E) \, dE = \phi_{th}$$

$$\int_{th} n(E) \, dE = \frac{\phi_{th}}{v_{av}}$$

and

$$\int_{th} \Sigma_{ac}(E) \phi(E) \, dE = \bar{\Sigma}_{ac} \phi_{th}$$

The average diffusion coefficient for thermal neutrons is defined as:

$$\bar{D}_c = \frac{\int_{th} D_c(E) \phi(E) \, dE}{\phi_{th}}$$

Now equation (4.46) becomes:

$$\frac{1}{v_{av}} \frac{\partial \phi_{th}(\mathbf{r}, t)}{\partial t} = \bar{D}_c \nabla^2 \phi_{th}(\mathbf{r}, t) - \bar{\Sigma}_{ac} \phi_{th}(\mathbf{r}, t) + S_{th}(\mathbf{r}, t) \quad (4.47)$$

The average absorption cross-section $\bar{\Sigma}_{ac}$ and the average diffusion coefficient $\bar{D}_c$ refer to the mixture of materials in the core of the reactor, as the suffix c is intended to emphasise. The absorption cross-section depends strongly on the fuel and its concentration in the reactor. On the other hand, particularly for dilute fuel–moderator mixtures, the diffusion coefficient (which is determined principally by the scattering cross-section) depends almost entirely on the moderator, and its value can be taken as nearly equal to the value for pure moderator.

The source of thermal neutrons per unit volume, $S_{th}(\mathbf{r})$, in a reactor is from fission. The rate of thermal fission is $\bar{\Sigma}_f \phi_{th}(\mathbf{r})$, the total number of fission neutrons produced from both thermal and fast fission is $\epsilon v \bar{\Sigma}_f \phi_{th}(\mathbf{r})$, and the number of neutrons which survive slowing down and become thermalized is $p\epsilon v \bar{\Sigma}_f \phi_{th}(\mathbf{r})$. (The leakage of neutrons while slowing down is neglected in this expression, but the capture of neutrons is taken into account by the factor p.)

The infinite reproduction constant may be expressed as:

$$k_\infty = \frac{\text{The rate of production of neutrons in the core}}{\text{The rate of absorption of neutrons in the core}}$$

$$= \frac{\int_{core} p \epsilon v \bar{\Sigma}_f \phi_{th}(\mathbf{r}) \, dV}{\int_{core} \bar{\Sigma}_{ac} \phi_{th}(\mathbf{r}) \, dV}$$

Since p, ϵ, v, and the cross-sections are independent of position in a homogeneous reactor, it follows from this equation that:

$$k_\infty \bar{\Sigma}_{ac} = \epsilon p v \bar{\Sigma}_f$$

and
$$S_{th}(\mathbf{r}) = k_\infty \bar{\Sigma}_{ac} \phi_{th}(\mathbf{r}) \quad (4.48)$$

Using this result, equation (4.47) becomes:

$$\frac{1}{v_{av}} \frac{\partial \phi_{th}(\mathbf{r}, t)}{\partial t} = \bar{D}_c \nabla^2 \phi_{th}(\mathbf{r}, t) - \bar{\Sigma}_{ac} \phi_{th}(\mathbf{r}, t) + k_\infty \bar{\Sigma}_{ac} \phi_{th}(\mathbf{r}, t)$$

or
$$\frac{1}{\bar{D}_c v_{av}} \frac{\partial \phi_{th}(\mathbf{r}, t)}{\partial t} = \nabla^2 \phi_{th}(\mathbf{r}, t) + \frac{k_\infty - 1}{L_c^2} \phi_{th}(\mathbf{r}, t) \quad (4.49)$$

where:
$$L_c^2 = \frac{\bar{D}_c}{\bar{\Sigma}_{ac}} \quad (4.50)$$

L_c^2 is the diffusion length squared of thermal neutrons in the fuel–moderator mixture that constitutes the core of the reactor.

At this point we define the material buckling B_m^2 of a reactor by the equation:

$$B_m^2 = \frac{k_\infty - 1}{L_c^2} \qquad (4.51)$$

and equation (4.49) becomes:

$$\frac{1}{\overline{D_c} v_{av}} \frac{\partial \phi_{th}(\mathbf{r}, t)}{\partial t} = \nabla^2 \phi_{th}(\mathbf{r}, t) + B_m^2 \phi_{th}(\mathbf{r}, t) \qquad (4.52)$$

If the thermal neutron flux $\phi_{th}(\mathbf{r}, t)$ is now expressed as the product of two functions, one of position $\mathbf{r}$ and one of time t, thus:

$$\phi_{th}(\mathbf{r}, t) = F(\mathbf{r})T(t)$$

equation (4.52) becomes:

$$\frac{1}{\overline{D_c} v_{av} T} \frac{dT}{dt} = \frac{\nabla^2 F}{F} + B_m^2 \qquad (4.53)$$

The material buckling B_m^2 is constant, therefore:

$$\frac{\nabla^2 F}{F} \text{ and } \frac{1}{T} \frac{dT}{dt}$$

are independently equal to constants. Let:

$$\frac{\nabla^2 F}{F} = -B_g^2$$

or

$$\nabla^2 F + B_g^2 F = 0 \qquad (4.54)$$

Equation (4.54) expresses the spatial variation of the flux in a reactor. Its solution subject to appropriate boundary conditions enables B_g^2 to be expressed in terms of the dimensions of a reactor, and the value of B_g^2 corresponding to the fundamental solution of this equation (which is the only solution of interest to us), is called the geometric buckling. (The name arises because B_g^2 is a measure of the curvature of the flux, or the extent to which it buckles.)

In a critical reactor, that is one in which $\partial \phi / \partial t$ (or dT/dt) is zero, it is clear, comparing equations (4.53) and (4.54), that:

$$B_m^2 = B_g^2 \qquad (4.55)$$

In a non-critical reactor B_m^2 is not equal to B_g^2, and if the reactor is supercritical, $\partial \phi / \partial t$ (or dT/dt) > 0, and $B_m^2 > B_g^2$. On the other hand if the reactor is subcritical, $\partial \phi / \partial t$ (or dT/dt) < 0, and $B_m^2 < B_g^2$.

In this chapter we are concerned only with critical reactors, so B_m^2 and B_g^2 can be regarded as always equal to each other and will be denoted simply by B^2, the buckling. For a critical reactor, equation (4.42) becomes:

$$B^2 = \frac{k_\infty - 1}{L_c^2} \qquad (4.56)$$

and this is the one-group critical equation. The reactor flux equation becomes:

$$\nabla^2\phi + B^2\phi = 0 \qquad (4.57)$$

(Henceforth in this chapter ϕ_{th} will be written simply as ϕ.) The fundamental solution of this equation together with equation (4.56) provides the necessary relationship between the materials of the core, (k_∞ and L_c^2), and its critical size, in which the buckling can be regarded as the link.

4.14 Diffusion length and slowing-down length

In the calculation of L_c^2 for a reactor the value of the average diffusion coefficient can, as already stated, usually be taken as being the same as the value for pure moderator. This is certainly the case for dilute fuel-moderator mixtures. The average absorption cross-section for the reactor is calculated by the method shown in Chapter 2, and is:

$$\bar\Sigma_{ac} = 0.8862 \sum_i N_i\sigma_{ai}$$

where N_i is the number of atoms/cm^3 of the reactor of each constituent i whose 2200 m/s absorption cross-section is σ_{ai}. These constituents will include the fuel and moderator, and possibly also coolant, cladding and structural materials.

If the core of the reactor can be regarded as consisting only of fuel and moderator, then, recalling equation (4.4):

$$f = \frac{\bar\Sigma_{aF}}{\bar\Sigma_{ac}} = \frac{\bar\Sigma_{aF}}{\bar\Sigma_{aF} + \bar\Sigma_{aM}}$$

and

$$1 - f = \frac{\bar\Sigma_{aM}}{\bar\Sigma_{ac}}$$

where $\bar\Sigma_{aM}$ is the average absorption cross-section of the moderator component of the reactor, $N_M\bar\sigma_{cM}$ (N_M being the number of atoms of moderator/cm^3 of the reactor). If the fuel-moderator mixture is dilute, then the volume occupied by fuel in the core is very small, and (in terms of volume) the core is nearly all moderator. Thus $\bar\Sigma_{aM}$ is very nearly equal to $\bar\Sigma_a$ for pure moderator and:

$$\bar\Sigma_{ac} = \frac{\bar\Sigma_a \text{ for pure moderator}}{1 - f}$$

Combining this result with the earlier statement regarding the average diffusion coefficient, we conclude that:

$$L_c^2 = (1 - f) \times L^2 \text{ for pure moderator} \qquad (4.58)$$

Thus the diffusion length for the reactor can be expressed in terms of the diffusion length of its moderator and the thermal utilization factor. Values of L^2 for the principal moderators are shown in Table 4.2. From these values it can be seen that L^2 for water is very much less than for heavy water and graphite, and thus the average crow flight distance travelled by thermal neutrons in a water moderated reactor is much less than in heavy water or graphite moderated reactors. This in turn means that water moderated reactors can be designed with much smaller cores than other reactor types. (The values of L^2 for graphite and heavy water depend on the purity of these materials, and the values given are for reactor grade materials).

Table 4.2. Diffusion and slowing-down constants for moderators

Material	$\bar{D}$ (cm)	L^2 (cm²)	L_s^2 (cm²)
Water	0·16	8·1	27
Heavy water	0·87	$3·0 \times 10^4$	131
Graphite	0·84	2650	368
Beryllium	0·50	480	102

Hitherto the distance travelled by neutrons during slowing down has been neglected; however, by analogy with the diffusion length squared, L^2 (which can be thought of as the product of the diffusion coefficient and the mean free path of thermal neutrons up to their point of absorption), the slowing-down length squared, L_s^2, is the diffusion coefficient multiplied by the mean total distance travelled by fission neutrons during slowing down to the point where they become thermalized. Thus:

$$L_s^2 = D \times \lambda_s \times \left(\begin{array}{c} \text{The average number of collisions} \\ \text{to thermalize fission neutrons} \end{array} \right)$$

$$= \frac{D}{\xi \Sigma_s} \log_e \frac{E_f}{E_{th}}$$

Using the same arguments that have been applied to D and Σ_s, it may be said that the value of L_s^2 for a dilute fuel-moderator mixture is very nearly equal to the value of L_s^2 for the pure moderator. These values are shown in Table 4.2 for the principal moderators.

It is possible to show that L_s^2 is equal to one-sixth of the mean square crow-flight distance travelled by neutrons during slowing down from fission to thermal energy, i.e.:

$$L_s^2 = \frac{1}{6} \overline{r_s^2} \tag{4.59}$$

where r_s denotes the crow-flight distance travelled by a neutron during slowing down. The similarity between this result and the result expressed by equation (4.45) for the diffusion length squared should be noted. L_s^2 has the same significance for the slowing down process as L^2 has for the diffusion of thermal neutrons.

Looking at the values of L_s^2 in Table 4.2, it is seen that water has the smallest value, which is not surprising as from the point of view of slowing neutrons down by scattering collisions, water has already been seen to be the most effective moderator. Thus water is seen to have the advantage over other moderators that both during slowing down and during thermal diffusion neutrons travel shorter distances than in other moderators, and the probability of neutron leakage from water moderated cores is less than from other cores of the same size.

4.15 Migration length and the modified one-group equation

The migration length squared M^2 for a reactor is defined as:

$$M^2 = L_c^2 + L_s^2 \qquad (4.60)$$

and its significance can be interpreted by reference to Figure 4.7 which shows the zig-zag track of a single neutron slowing down from fission to thermal energy and diffusing at this energy until it is absorbed.

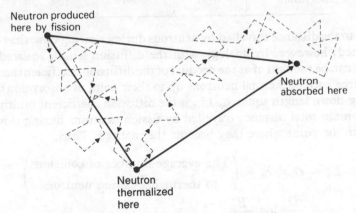

Figure 4.7. Neutron slowing down and diffusion

The crow-flight distance travelled during slowing down is represented by the vector $\mathbf{r}_s$, and the crow-flight distance travelled during thermal diffusion is represented by the vector $\mathbf{r}$. The total crow-flight distance travelled by the neutron from production to absorption is represented by the vector $\mathbf{r}_t$. If the angle between the vectors $\mathbf{r}_r$ and $\mathbf{r}$ is 90°, then $r_t^2 = r_s^2 + r^2$. Now thinking in terms of a large number of neutrons, if scattering is assumed to be isotropic then the average angle between the vectors $\mathbf{r}_s$ and $\mathbf{r}$ is 90°, and the average values of r_s^2, r^2 and r_t^2 are related by:

$$\overline{r_t^2} = \overline{r_s^2} + \overline{r^2}$$

Comparing this equation with (4.60), and recalling also equations (4.45) and (4.59), it follows that M^2 can be interpreted as one-sixth of the mean square crow-flight distance travelled by neutrons from the point where they are produced by fission to the point where they are ultimately absorbed after thermalization. Thus M^2 has the same significance for the total crow-flight distance travelled by neutrons as L^2 has for the crow-flight distance travelled after they become thermalized.

In order to take account of the diffusion of neutrons during slowing down, which was ignored in the analysis leading to the one-group equation, L_c^2 in the one-group critical equation (4.56) can be replaced by M^2 to give the modified one-group critical equation:

$$B^2 = \frac{k_\infty - 1}{M^2} \tag{4.61}$$

This equation has the advantages of being much more accurate than the one-group equation, while at the same time being no more difficult to use. Its use is therefore preferred to the one-group equation.

4.16 Solution of the reactor equation

The reactor equation, (4.57), will now be solved for a bare cylindrical reactor core, the most common shape. The corresponding results for the other possible core shapes are tabulated at the end of this section. Boundary condition 4 can be applied at the surface of a bare core as it is assumed that no neutrons which leak from the core can return to it.

Figure 4.8 shows a bare cylindrical core whose dimensions are height L and radius R. If the extrapolation distance is $0.71\lambda_{tr}$, the extrapolated dimensions of the core are given by the equations:

$$L' = L + 1.42\lambda_{tr}$$

and
$$R' = R + 0.71\lambda_{tr}$$

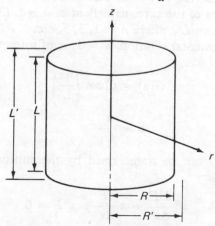

Figure 4.8. Bare cylindrical reactor core

In cylindrical coordinates with angular symmetry the reactor equation is:

$$\frac{\partial^2 \phi}{\partial r^2} + \frac{1}{r} \frac{\partial \phi}{\partial r} + \frac{\partial^2 \phi}{\partial z^2} + B^2 \phi = 0 \tag{4.62}$$

It is assumed that the flux $\phi(r, z)$ can be expressed as the product of two functions, one of r only and the other of z only, thus:

$$\phi(r, z) = F(r)G(z)$$

Equation (4.62) can now be written as:

$$\frac{1}{F}\frac{d^2F}{dr^2} + \frac{1}{Fr}\frac{dF}{dr} + \frac{1}{G}\frac{d^2G}{dz^2} + B^2 = 0 \qquad (4.63)$$

$\left\{\dfrac{1}{F}\dfrac{d^2F}{dr^2} + \dfrac{1}{Fr}\dfrac{dF}{dr}\right\}$ and $\dfrac{1}{G}\dfrac{d^2G}{dz^2}$ are independent of each other, and each is equal to some constant. Let:

$$\frac{1}{F}\frac{d^2F}{dr^2} + \frac{1}{Fr}\frac{dF}{dr} = -\alpha^2 \qquad (4.64)$$

and

$$\frac{1}{G}\frac{d^2G}{dz^2} = -\beta^2 \qquad (4.65)$$

from equations (4.63), (4.64) and (4.65) we see that:

$$\alpha^2 + \beta^2 = B^2 \qquad (4.66)$$

The solution of equation (4.65) is:

$$G(z) = A \sin \beta z + C \cos \beta z$$

The solution must be symmetrical above and below the centre plane of the core, $z = 0$, therefore $A = 0$. The flux must be zero at the extrapolated boundaries of the core, namely at $z = \pm L'/2$. This condition is satisfied if $\beta = n\pi/L'$, where $n = 1, 3, 5$, etc.

Only the fundamental solution, $n = 1$, is of interest, thus the solution of equation (4.65) is:

$$G(z) = C \cos\left(\frac{\pi z}{L'}\right) \qquad (4.67)$$

and

$$\beta = \frac{\pi}{L'} \qquad (4.68)$$

Equation (4.64) can be transformed by the substitution $x = \alpha r$ to give:

$$x^2\frac{d^2F}{dx^2} + x\frac{dF}{dx} + x^2F = 0 \qquad (4.69)$$

This is Bessel's ordinary equation of zero order, of which the solution is:

$$F(x) = D\mathcal{J}_0(x) + EY_0(x)$$

or

$$F(r) = D\mathcal{J}_0(\alpha r) + EY_0(\alpha r)$$

where $\mathcal{J}_0$ and Y_0 are ordinary Bessel functions of zero order. These functions are shown in Figure 4.9.

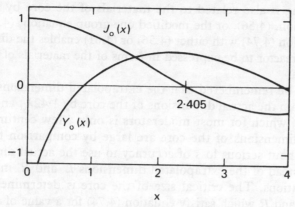

Figure 4.9. Ordinary Bessel functions of zero order

As $\alpha r \to 0$, $Y_0(\alpha r) \to -\infty$, which would give a physically impossible solution, therefore $E = 0$. The flux must be zero at the extrapolated radius of the core, R', and considering the fundamental solution of the equation $F(r) = D\mathcal{J}_0(\alpha r)$, the result for α is $\alpha = 2\cdot405/R'$ since the first zero of the $\mathcal{J}_0$ function is at $2\cdot405$. The solution of equation (4.64) is therefore:

$$F(r) = D\mathcal{J}_0\left(\frac{2\cdot405r}{R'}\right) \qquad (4.70)$$

and

$$\alpha = \frac{2\cdot405}{R'} \qquad (4.71)$$

Combining equations (4.67) and (4.70), the complete solution for the flux in a cylindrical reactor core is:

$$\phi(r, z) = A' \cos\left(\frac{\pi z}{L'}\right)\mathcal{J}_0\left(\frac{2\cdot405r}{R'}\right) \qquad (4.72)$$

where $A' = C \times D$, and represents the value of the flux at the centre of the core where $r = z = 0$. This is the maximum value of the flux, and the preceding equation may be written:

$$\phi(r, z) = \phi_{max} \cos\left(\frac{\pi z}{L'}\right)\mathcal{J}_0\left(\frac{2\cdot405r}{R'}\right) \qquad (4.73)$$

This equation expresses the flux variation in the core, and the buckling B^2 is related to the dimensions of the core by using equations (4.66), (4.68) and (4.71) to give:

$$B^2 = \left(\frac{\pi}{L'}\right)^2 + \left(\frac{2\cdot405}{R'}\right)^2 \qquad (4.74)$$

The buckling is also related to the materials of the core by the one-group equation, (4.56), or the modified one-group equation, (4.61). The use of equation (4.74) with either (4.56) or (4.61) enables the dimensions of a critical reactor to be expressed in terms of the materials of the core, or vice versa.

It should be remembered that the extrapolated dimensions, L' and R', differ from the actual dimensions of the core by $1 \cdot 42\lambda_{tr}$ and $0 \cdot 71\lambda_{tr}$ respectively, which for most moderators is only a few centimetres. If the actual dimensions of the core are large by comparison then it is possible without serious loss of accuracy to use the actual dimensions L and R instead of the extrapolated dimensions L' and R' in the preceding equations. The critical size of the core is determined by the values of L and R which satisfy equation (4.74) for a value of B^2 which is determined by the materials of the reactor.

The solution of the reactor equation for the other core shapes follows the same procedure as outlined above for a cylindrical core, and the results are summarized below in Table 4.3.

Table 4.3. The solution of the reactor equation for the principal core geometries

Shape	Extrapolated dimensions	Flux	Buckling B^2
Rectangular parallelepiped	$a' \times b' \times c'$	$\phi = \phi_{max} \cos\left(\dfrac{\pi x}{a'}\right) \cos\left(\dfrac{\pi y}{b'}\right) \cos\left(\dfrac{\pi z}{c'}\right)$	$\left(\dfrac{\pi}{a'}\right)^2 + \left(\dfrac{\pi}{b'}\right)^2 + \left(\dfrac{\pi}{c'}\right)^2$
Cylinder	$L' \times R'$	$\phi = \phi_{max} \cos\left(\dfrac{\pi z}{L'}\right) \mathcal{J}_0\left(\dfrac{2 \cdot 405 r}{R'}\right)$	$\left(\dfrac{\pi}{L'}\right)^2 + \left(\dfrac{2 \cdot 405}{R'}\right)^2$
Sphere	R'	$\phi = \phi_{max} \dfrac{R'}{\pi r} \sin\left(\dfrac{\pi r}{R'}\right)$	$\left(\dfrac{\pi}{R'}\right)^2$

The question arises in the case of rectangular parallelepiped and cylindrical reactors as to the values of the dimensions which, for a fixed composition and value of B^2, give the minimum critical volume, and hence the minimum mass of fuel in the core. The problem can be solved easily for a cylindrical core, neglecting the difference between the actual and the extrapolated dimensions. The conditions are:

$$B^2 = \left(\frac{\pi}{L}\right)^2 + \left(\frac{2 \cdot 405}{R}\right)^2$$

is constant, and the volume, $\pi R^2 L$ is to be minimized. The volume can be expressed as:

$$V = \frac{\pi L \times 2 \cdot 405^2}{\{B^2 - (\pi/L)^2\}}$$

and is a minimum when dV/dL is zero, which gives the result:

$$L = \frac{\sqrt{3}\,\pi}{B} \tag{4.75}$$

From this it follows that:

$$R = \sqrt{\frac{3}{2}}\,\frac{(2\cdot405)}{B} \quad \text{and} \tag{4.76}$$

$$V = \frac{148}{B^3} \tag{4.77}$$

The ratio of the radius to the height is given by:

$$\frac{R}{L} = \frac{2\cdot405}{\sqrt{2}\,\pi} = 0\cdot54 \tag{4.78}$$

Thus the minimum volume of a cylindrical core of fixed buckling is obtained when the diameter is approximately equal to the height. Figure 4.10 shows the variation of R with L for a cylindrical core of fixed buckling, and also the variation of the critical mass with L. The latter curve shows a minimum point which corresponds to the optimum dimensions just derived.

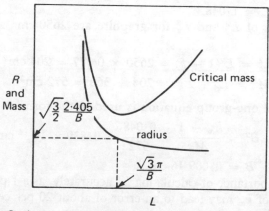

Figure 4.10. Optimum dimensions and critical mass of a cylindrical core

In the case of a rectangular parallelepiped it can be shown that the minimum volume, as might be expected, occurs when the three sides are equal and the shape of the core is a cube. Table 4.4 shows the optimum dimensions and the corresponding volume of the three core shapes in terms of the buckling. The minimum volume is obtained with a spherical core, which is to be expected as of the three core shapes the sphere has, for a given volume, the minimum surface area. Since neutron leakage is a surface effect, the leakage per unit volume of core is least for a sphere, and this is therefore from a purely physical point of view the best core shape. From the practical point of view, however, it has disadvantages and is rarely used.

Table 4.4. Optimum core dimensions and minimum volumes

Core shape	Optimum dimensions	Minimum volume
Cube	$a = b = c = \dfrac{\sqrt{3}\,\pi}{B}$	$\dfrac{161}{B^3}$
Cylinder	$L = \dfrac{\sqrt{3}\,\pi}{B}, R = \dfrac{2 \cdot 405}{B}\sqrt{\dfrac{3}{2}}$	$\dfrac{148}{B^3}$
Sphere	$R = \dfrac{\pi}{B}$	$\dfrac{130}{B^3}$

Example: Calculate the dimensions of the cylindrical core and the minimum critical mass of uranium for a homogeneous reactor consisting of a mixture of enriched uranium (2 per cent ^{235}U) and graphite in the ratio 1 to 300 by atoms.

The necessary cross-sections, densities and atomic weights are tabulated in Appendix 2.

The four factors of k_∞ are calculated by the same method as illustrated in the example earlier in this chapter. The values are (see Table 4.1):

$$\epsilon = 1 \cdot 00, \quad p = 0 \cdot 659, \quad f = 0 \cdot 923, \quad \eta = 1 \cdot 722$$

from which $k_\infty = 1 \cdot 048$.

The values of L^2 and L_s^2 for graphite are 2650 cm^2 and 368 cm^2 respectively.

$$L_c^2 = L^2(1 - f) = 2650 \times 0 \cdot 077 = 204 \text{ cm}^2$$
$$M^2 = L_c^2 + L_s^2 = 204 + 368 = 572 \text{ cm}^2$$

The modified one-group equation is used in the form

$$B^2 = \frac{k_\infty - 1}{M^2} = \frac{0 \cdot 048}{572} = 0 \cdot 839 \times 10^{-4} \text{ cm}^{-2}$$
$$B = 0 \cdot 009 \, 16 \text{ cm}^{-1}$$

(Note the importance of calculating k_∞ accurately, as a 1 per cent error in the value of k_∞ may lead to an error of about 20 per cent in $k_\infty - 1$, and hence B^2.)

The minimum core volume, and hence minimum critical mass of uranium is obtained with a radius to height ratio of 0·54. From Table 4.4, neglecting extrapolation distances, the dimensions of the core are $L = 5 \cdot 94$ m, $R = 3 \cdot 22$ m, and the volume = 193 m^3.

The mass of uranium is calculated by neglecting the fractional volume of fuel in the core and calculating first the number of atoms of graphite per cm^3 of the core.

$$N(C) = \frac{1620 \times 6 \cdot 023 \times 10^{26}}{12} = 8 \cdot 13 \times 10^{28} \text{ atoms/m}^3$$

where the density of graphite has been taken as 1620 kg/m^3, and its atomic mass is 12.

The number of atoms of uranium per m^3 of the core:

$$N(U) = \frac{8 \cdot 13 \times 10^{28}}{300} = 2 \cdot 71 \times 10^{26} \text{ atoms/m}^3$$

The mass of uranium per m^3 of the core:

$$M(U) = \frac{2 \cdot 71 \times 10^{26} \times 238}{6 \cdot 023 \times 10^{26}} = 107 \text{ kg/m}^3$$

The mass of uranium in the reactor $= 107 \times 193 \times 10^{-3}$

$$= 20 \cdot 7 \text{ tonnes}$$

The error involved by neglecting the extrapolation distances may be estimated. The extrapolation distance for a graphite moderated reactor is about 2 cm, and the error involved in the length and radius of the core by neglecting it is about $0 \cdot 7$ per cent. The error in the volume of the core and the critical mass of uranium is therefore about 2 per cent.

4.17 Neutron leakage

In Chapter 3 (section 3.2) the on-leakage probabilities for fast and thermal neutrons, P_{NLf} and P_{NLth}, were defined. The fraction of all fission produced neutrons which do not leak out of the core, P_{NL}, is related to these two non-leakage probabilities by:

$$P_{\text{NL}} = P_{\text{NLth}} \times P_{\text{NLf}}$$

and the fraction of all fission produced neutrons which do leak out of the core, P_{L}, is given by:

$$P_{\text{L}} = 1 - P_{\text{NL}} = 1 - (P_{\text{NLth}} \times P_{\text{NLf}})$$

In a critical finite reactor in which the value of $k_{\text{eff}} = 1$, equation (3.3) shows that k_∞ and P_{L} are related by:

$$P_{\text{L}} = 1 - \frac{1}{k_\infty} \tag{4.79}$$

According to one-group theory in which all neutrons are assumed to be thermal neutrons, the leakage of fast neutrons is neglected and therefore $P_{\text{L}} = P_{\text{Lth}}$. Using equation (4.56) in the form $k_\infty = 1 + B^2 L_c^2$ with equation (4.79) it can be shown that the leakage of thermal neutrons is:

$$P_{\text{Lth}} = \frac{B^2 L_c^2}{1 + B^2 L_c^2} \tag{4.80}$$

According to modified one-group theory, equation (4.61) in the form $k_\infty = 1 + B^2 M^2$ can be combined with equation (4.79) to show that the leakage fraction of all neutrons, fast and thermal, is given by:

$$P_{\text{L}} = \frac{B^2 M^2}{1 + B^2 M^2} \tag{4.81}$$

4.18 Form factors

As will become clear in a later chapter when we consider the rate of energy release and heat transfer in a power reactor, it is important to know the relationship between the maximum rate of energy release and the average rate of energy release in the core. These quantities are in turn proportional to the maximum and average thermal neutron fluxes in the core.

The rate of energy release per unit volume at any point in a reactor core is given by:

$$H(\mathbf{r}) = 3{\cdot}2 \times 10^{-11}\, \overline{\Sigma}_f \phi_{th}(\mathbf{r})\ \text{W/cm}^3$$

The thermal output of the whole reactor, Q_r, is:

$$Q_r = 3{\cdot}2 \times 10^{-11} \int_{core} \overline{\Sigma}_f \phi_{th}(\mathbf{r})\ \mathrm{d}V \text{ watts,}$$

or alternatively

$$Q_r = 3{\cdot}2 \times 10^{-11} \times \overline{\Sigma}_f \times \text{Volume of core} \times \phi_{av}$$

where ϕ_{av} is the spatially averaged thermal neutron flux in the core.

In a cylindrical reactor core the relationship between the average and maximum fluxes can be derived from equation (4.73) as:

$$\phi_{av} = \frac{1}{\pi R^2 L} \int_0^R \int_{-L/2}^{L/2} \phi_{max} \cos\left(\frac{\pi z}{L'}\right) \mathcal{J}_0\left(\frac{2{\cdot}405 r}{R'}\right) 2\pi r\ \mathrm{d}r\ \mathrm{d}z \quad (4.82)$$

It is convenient to define axial (α_z) and radial (α_r) form factors for a cylindrical reactor as follows:

$$\alpha_z = \frac{\text{The flux at the centre of the core}}{\text{The average flux along the central axis of the core}}$$

$$= \frac{\phi_{max}}{(1/L) \int_{-L/2}^{L/2} \phi_{max} \cos(\pi z/L')\ \mathrm{d}z}$$

$$= \frac{\pi L/2L'}{\sin(\pi L/2L')} \quad (4.83)$$

If $L = L'$, that is if the extrapolation distance is neglected, then:

$$\alpha_z = \pi/2 = 1{\cdot}57.$$

$$\alpha_r = \frac{\text{The flux at the centre of the core}}{\text{The average flux in the central plane of the core at } z = 0}$$

$$= \frac{\phi_{max}}{(1/\pi R^2) \int_0^R \phi_{max} \mathcal{J}_0(2{\cdot}405 r/R') 2\pi r\ \mathrm{d}r}$$

$\int r \mathcal{J}_0(\beta r)\ \mathrm{d}r = (1/\beta) r \mathcal{J}_1(\beta r)$, and the values of $\mathcal{J}_1(\beta r)$ at $\beta r = 0$ and $2{\cdot}405$ are 0 and $0{\cdot}52$ respectively. The expression for α_r becomes:

$$\alpha_r = \frac{1 \cdot 2R/R'}{\mathscr{J}_1(2 \cdot 405R/R')} \tag{4.84}$$

and if the extrapolation distances are neglected, i.e. if $R = R'$:

$$\alpha_r = 2 \cdot 31$$

The overall form factor α_0 is the product of the axial and radial form factors, and using the preceding equations for α_z and α_r it is seen that the overall form factor is the ratio of the maximum to average flux in the core:

$$\alpha_0 = \alpha_z \times \alpha_r = \frac{\phi_{max}}{\phi_{av}} \tag{4.85}$$

If extrapolation distances are neglected, $\alpha_0 = 3 \cdot 63$.

4.19 Conclusion

The accuracy of any calculations based on the theory outlined in this chapter depends on two factors—the accuracy of the theory itself, and the accuracy with which data such as cross-sections and the value of ν are known. The limitations of one-group theory have already been mentioned, and it should be pointed out that cross-sections and other nuclear data, which must be determined experimentally, are not known with complete accuracy. Consequently calculations alone are not sufficient for the design of reactors and the exact prediction of criticality, and experimental results also have to be used. One example of the use of experimental data is in the empirical formula for the effective resonance integral which was introduced earlier in this chapter, and two further examples will be mentioned. Both these examples depend on the fact that if in a sub-critical assembly (one that is either too small or has too low a concentration of fuel) a source of neutrons is present, a steady-state flux is established in the system, and this flux may be measured.

In the exponential experiment an assembly of fuel and moderator is built of a size considerably less than the critical size, and a source of neutrons is located at one side of the assembly. Measurements of the flux distribution in the assembly enable the value of k_∞ to be determined. In the "approach to critical" experiment the reactor core is constructed with a neutron source, and the fuel is loaded progressively, one fuel element at a time, into the core. The variation of the neutron flux as the fuel is loaded enables the critical mass of fuel to be very accurately predicted before it is actually loaded into the core and the reactor becomes critical.

Finally, it should be pointed out that it is not sufficient to design a reactor whose initial value of k_{eff} is exactly 1. If this were done, it would not be possible, without a neutron source, to start up the reactor, and once the reactor is started up it would soon become subcritical and shut down as some fissile material is used up and other changes occur. A reactor is always designed so that the initial value of k_{eff} is slightly greater than 1. The excess reactivity δk, which is defined as $k_{eff} - 1$, may be in the range 0·01 to 0·25 or more depending on the type of reactor, and it is offset at the start of the reactor's life by inserting into the core rods of neutron absorbing material known as control rods. These rods are progressively withdrawn to keep the value of k_{eff} equal to 1 and enable the reactor to operate at steady power for long periods of time. This subject will be dealt with in more detail in a later chapter, however the point to note at this stage is that the method developed in this chapter may be modified slightly to determine the composition or size of a reactor with a certain amount of excess reactivity.

For example, if it is desired to find the size of a reactor of given composition and a given excess reactivity, δk, the value of k_∞ for the reactor is calculated in the normal way by the four factor formula, and L_c^2 and L_s^2 are obtained. Equation (4.61) is then used in the modified form

$$B^2 = \frac{(k_\infty - \delta k) - 1}{M^2} \tag{4.86}$$

to determine B^2, and hence the dimensions of the core. This is equivalent to using a fictitious value of the infinite multiplication factor which is equal to the actual value minus the excess reactivity.

Chapter 5

The theory of nuclear reactors— further topics

In the previous chapter we developed the one-group theory of bare homogeneous reactors to the point where it was possible to make simple calculations of critical size and critical mass of fuel for a reactor. In this chapter we will introduce some further topics in nuclear reactor theory to give a more complete picture of the subject.

Of the various reactor models mentioned in the introduction to the last chapter only the one-group model was developed fully, although the continuous slowing down model was used to study resonance capture. In this chapter we will study very briefly the two-group model, and use the continuous slowing-down model to analyse the combined slowing down and diffusion of neutrons. These two models yield slightly different forms of the critical equation for bare reactors, which can be compared with the one-group and modified one-group equations.

The theory of homogeneous reactors is to some extent limited in its application, and although many reactors can be regarded as quasi-homogeneous there are also many that can not, and these must be analysed by methods that take account of their heterogeneous structure. Furthermore few reactors are bare, and there are obvious advantages to be gained by surrounding the core by some material which scatters neutrons, some of which return into the core, thereby reducing leakage and the critical mass of fuel. Such a material is called a reflector, and the reactor is described as a reflected reactor.

The theories of heterogeneous reactors and reflected reactors are much more complicated than the theory of bare homogeneous reactors. For this reason this chapter will only present some qualitative descriptions and very simple models to point out the principal physical differences between these reactors.

5.1 Two-group theory for bare reactors

For accurate design work it is not sufficient to consider all neutrons in one energy group as in one-group theory, and in multi-group models neutrons are subdivided according to their energy into several groups, each one representing a particular energy range. Diffusion equations for each group are set up using appropriate constants, and the simultaneous solution of these equations provides the condition for criticality of the reactor.

The method will be demonstrated by considering the two-group model, which has the advantages that it is considerably more accurate than the one-group model, and at the same time is the simplest of the multi-group models and the only one that does not require a computer for the numerical solution of problems.

In the two-group model all thermal neutrons are considered to be in one group, the thermal group, and all neutrons slowing down from fission to thermal energy are considered to be in the other group, called the fast group. Neutrons in the fast group remain in that group until they are captured, leak out of the core, or undergo a sufficient number of scattering collisions to become thermalized and enter the thermal group.

For fission neutrons to be scattered from the fast to the thermal group, they must each suffer on the average $[\log (E_f/E_{th})]/\xi$ elastic scattering collisions. The effective cross-section Σ_1 for the removal of neutrons from the fast group is therefore given, in the absence of resonance capture, by:

$$\Sigma_1 = \frac{\Sigma_s \xi}{\log (E_f/E_{th})} \tag{5.1}$$

where Σ_s is the scattering cross-section for fast neutrons.

From the definition of the slowing-down length given in the last chapter it is evident that:

$$\Sigma_1 = \frac{D_1}{L_s^2} \tag{5.2}$$

where D_1 is the diffusion coefficient of fast neutrons.

The steady-state diffusion equations for the two groups of neutrons are written as:

(Fast group) $D_1 \nabla^2 \phi_1 - \Sigma_1 \phi_1 + \epsilon \nu \Sigma_{f2} \phi_2 = 0$ (5.3)

and (Thermal group) $D_2 \nabla^2 \phi_2 - \Sigma_{a2} \phi_2 + p \Sigma_1 \phi_1 = 0$ (5.4)

(Suffix 1 refers to the fast group, and suffix 2 refers to the thermal group.) Note that the source of neutrons in the fast group (from thermal and fast fission) is given by $\epsilon \nu \Sigma_{f2} \phi_2$, and that resonance capture has been taken into account by expressing the rate at which fast neutrons enter the thermal group as $p\Sigma_1\phi_1$.

The homogeneous parts of equations (5.3) and (5.4) are similar to the one-group equation $\nabla^2\phi + B^2\phi = 0$, so we assume solutions of the same form, i.e.:

$$\nabla^2\phi_1 = -B_1^2\phi_1$$

and

$$\nabla^2\phi_2 = -B_2^2\phi_2$$

If these expressions for $\nabla^2\phi_1$ and $\nabla^2\phi_2$ are substituted into equations (5.3) and (5.4) it can be shown that $\phi_1 = \phi_2 \times$ a constant determined by the material properties of the core. From this it follows that $\nabla^2\phi_1 = \nabla^2\phi_2 \times$ the same constant, therefore:

$$\frac{\nabla^2\phi_1}{\phi_1} = \frac{\nabla^2\phi_2}{\phi_2} \tag{5.5}$$

or

$$B_1^2 = B_2^2 \tag{5.6}$$

In other words the solutions B_1^2 and B_2^2 assumed above are the same, and will be denoted by B^2, the buckling of both fast and thermal fluxes in the core. Replacing $\nabla^2\phi_1$ and $\nabla^2\phi_2$ by $-B^2\phi_1$ and $-B^2\phi_2$, and noting that:

$$k_\infty = \frac{\epsilon p\nu\Sigma_{f2}}{\Sigma_{a2}} \tag{5.7}$$

equations (5.3) and (5.4) become:

$$(1 + B^2L_s^2)\phi_1 - \frac{k_\infty\Sigma_{a2}}{p\Sigma_1}\phi_2 = 0 \tag{5.8}$$

and

$$(1 + B^2L_c^2)\phi_2 - \frac{p\Sigma_1}{\Sigma_{a2}}\phi_1 = 0 \tag{5.9}$$

where L_c^2 is the diffusion length squared of thermal neutrons in the core of the reactor. The solution of these equations is non-trivial provided that the determinant

$$\begin{vmatrix} (1 + B^2L_s^2) & -\dfrac{k_\infty\Sigma_{a2}}{p\Sigma_1} \\[2ex] -\dfrac{p\Sigma_1}{\Sigma_{a2}} & (1 + B^2L_c^2) \end{vmatrix} = 0$$

or

$$k_\infty = (1 + B^2L_s^2)(1 + B^2L_c^2) \tag{5.10}$$

This is the critical equation of a bare reactor according to two-group theory, and it should be compared with the one-group and modified one-group equations, (4.56) and (4.61). The similarity between these equations is obvious; in particular, if the right-hand side of equation (5.10) is expanded and the term involving B^4 is neglected (it will be very small for a large core), this equation becomes identical with the modified one-group equation, (4.61). Equation (5.10) is a quadratic in B^2; however, of the two values of B^2 which satisfy it, one is negative and does not contribute to the solution for a bare core. The positive value of B^2 is given by:

$$B^2 = \frac{1}{2L_s^2 L_c^2} [-(L_s^2 + L_c^2) + \sqrt{(L_s^2 + L_c^2)^2 + 4(k_\infty - 1)L_s^2 L_c^2}]$$

A simpler approximate form is:

$$B^2 \approx \frac{k_\infty - 1}{L_s^2 + L_c^2} \qquad (5.11)$$

and the resemblance to the modified one-group equation is again evident.

The leakage probabilities for fast (P_{Lf}) and thermal (P_{Lth}) neutrons can be inferred from equations (3.3) and (5.10) as:

$$P_{Lf} = \frac{B^2 L_s^2}{1 + B^2 L_s^2} \qquad (5.12)$$

$$P_{Lth} = \frac{B^2 L_c^2}{1 + B^2 L_c^2} \qquad (5.13)$$

5.2 Fermi age theory

Another modification to one-group theory can be made to take into account the diffusion of fast neutrons while slowing down. This modification involves the use of the continuous slowing-down model, which was introduced in the previous chapter, and is known as Fermi Age theory.

A new variable called the neutron age, τ, will first be introduced. It may be defined as follows:

The age, τ, corresponding to energy E, is:

$$\tau(E) = \int_E^{E_1} \frac{D(E)}{\xi \Sigma_t(E)} \frac{\mathrm{d}E}{E} \qquad (5.14)$$

where E_1 is the source energy of the neutrons (2 MeV in the case of fission produced neutrons). From equation (5.14) it is clear that the age of fission neutrons is zero, and that the age increases as neutrons lose energy.

In a medium in which fission neutrons are slowing down and being thermalized, i.e. a medium in which $\Sigma_s \gg \Sigma_a$, or $\Sigma_s \approx \Sigma_t$, the age of thermal neutrons can be shown, using equation (5.14) and assuming that D, Σ_s and ξ are independent of energy, to be given by:

$$\tau_{th} = \frac{D}{\Sigma_s} \frac{\log E_f - \log E_{th}}{\xi} \qquad (5.15)$$

This equation may be interpreted as:

$\tau_{th} = D \times$ (The scattering mean free path) $\times$ (The average number of collisions required to thermalize fission neutrons).

$\phantom{\tau_{th}} = D \times$ (The mean total distance travelled by fission neutrons during slowing down to the point where they become thermalized).

Evidently τ_{th} is the same quantity as L_s^2 which was introduced in the last chapter to characterize the distance travelled by neutrons during slowing down. The names "age" and "slowing-down length squared" can both be used for this quantity. Despite its name, τ has the dimensions of length squared, and it can be shown by considering neutron slowing down in a non-absorbing medium that:

$$\tau_{th} = \tfrac{1}{6}\overline{r_s^2} \qquad (5.16)$$

where r_s is the crow-flight distance travelled by fission neutrons from their source to the point where they become thermalized. When this result is compared with equation (4.45), the similarity between τ_{th} and L^2 for fast and thermal neutrons respectively is emphasized.

The combined diffusion and slowing down of neutrons in the energy range E to $E + dE$, $(E_f > E > E_{th})$, is expressed by the steady-state diffusion equation in the form:

$$D(E)\, \nabla^2 \phi(\mathbf{r}, E)\, dE - \Sigma_a(E)\phi(\mathbf{r}, E)\, dE + S(\mathbf{r}, E)\, dE = 0 \qquad (5.17)$$

The source of neutrons in the energy interval dE at E is due to the difference in the rates at which neutrons slow down into and out of this interval:

$$S(\mathbf{r}, E)\, dE = q(\mathbf{r}, E + dE) - q(\mathbf{r}, E)$$

Expressing $q(\mathbf{r}, E + dE)$ in terms of $q(\mathbf{r}, E)$ by means of the first two terms of a Taylor series we get:

$$S(\mathbf{r}, E)\, dE = \frac{\partial q}{\partial E}(\mathbf{r}, E)\, dE,$$

and using this result in equation (5.17) we get:

$$D(E)\, \nabla^2 \phi(\mathbf{r}, E)\, dE - \Sigma_a(E)\phi(\mathbf{r}, E)\, dE + \frac{\partial q}{\partial E}(\mathbf{r}, E)\, dE = 0 \qquad (5.18)$$

Recalling a result of the continuous slowing-down model, namely:

$$\phi(E)\, dE = \frac{q(E)\, dE}{\xi \Sigma_t(E) E}$$

and assuming that the spatial variation of the slowing-down density is the same as the spatial variation of the neutron flux, i.e. $\nabla^2 q = \nabla^2 \phi$ at all energies, equation (5.18) becomes:

$$\frac{D(E)}{\xi\Sigma_t(E)} \nabla^2 q(\mathbf{r}, E) - \frac{\Sigma_a(E)}{\xi\Sigma_t(E)} q(\mathbf{r}, E) + E\frac{\partial q}{\partial E}(\mathbf{r}, E) = 0 \quad (5.19)$$

This equation is solved by assuming that the slowing-down density q, which is a function of position and energy, can be treated as the product of two functions, one of position only, and the other of energy only, thus:

$$q(\mathbf{r}, E) = F(\mathbf{r}) \times G(E)$$

Making use of this, equation (5.19) becomes, dividing throughout by $\frac{D(E)FG}{\xi\Sigma_t(E)}$:

$$\frac{\nabla^2 F(\mathbf{r})}{F(\mathbf{r})} - \frac{\Sigma_a(E)}{D(E)} + \frac{\xi\Sigma_t(E)E}{D(E)G(E)} \frac{dG(E)}{dE} = 0 \quad (5.20)$$

The first term on the left-hand side, $\nabla^2 F/F$, is a function of position only, and the other two terms are functions of energy only. It follows that:

$$\frac{\nabla^2 F(\mathbf{r})}{F(\mathbf{r})} = \frac{\Sigma_a(E)}{D(E)} - \frac{\xi\Sigma_t(E)E}{D(E)G(E)} \frac{dG(E)}{dE} = \text{a constant}$$

The spatial variation of F is, as already noted, the same as the spatial variation of ϕ, and the variation of ϕ is given by the equation, familiar from one-group theory:

$$\nabla^2 \phi = -B^2 \phi$$

or

$$\frac{\nabla^2 \phi}{\phi} = -B^2$$

It follows that:

$$\frac{\nabla^2 F}{F} = -B^2$$

where B^2 is the buckling already encountered in one-group theory.

The energy dependent part of equation (5.20) can now be written as:

$$\frac{dG(E)}{G(E)} = \frac{\Sigma_a(E)}{\xi\Sigma_t(E)} \frac{dE}{E} + \frac{B^2 D(E)}{\xi\Sigma_t(E)} \frac{dE}{E} \quad (5.21)$$

Integrating from fission to thermal energy, reversing the sign and taking antilogarithms we get:

$$\frac{G_{E_{th}}}{G_{E_f}} = \exp\left[-\int_{E_{th}}^{E_f} \frac{\Sigma_a(E)\,dE}{\xi\Sigma_t(E)\,E}\right] \times \exp\left[(-B^2)\int_{E_{th}}^{E_f} \frac{D(E)\,dE}{\xi\Sigma_t(E)E}\right] \quad (5.22)$$

The first term on the right-hand side is recognized as the resonance escape probability, and the second term may be simplified by using the definition of τ_{th}. Equation (5.22) becomes:

$$\frac{G_{E_{th}}}{G_{E_f}} = p \exp\left(-B^2\tau_{th}\right) \tag{5.23}$$

From which

$$\frac{q_{E_{th}}}{q_{E_f}} = p \exp\left(-B^2\tau_{th}\right) \tag{5.24}$$

$q_{E_{th}}$ is the slowing-down density at thermal energy which is the number of neutrons becoming thermal per cm^3 per second, and q_{E_f} is the slowing-down density at fission energy which is the number of neutrons per cm^3 per second which are produced by fission. The ratio $q_{E_{th}}/q_{E_f}$ is, by definition, the product of the resonance escape probability and the fast non-leakage probability. The term $\exp\left(-B^2\tau_{th}\right)$ can therefore be identified as the fast non-leakage probability.

Integrating equation (5.18) over the thermal neutron spectrum we get:

$$\bar{D}\nabla^2\phi_{th}(\mathbf{r}) - \bar{\Sigma}_a\phi_{th}(\mathbf{r}) + q_{E_{th}}(\mathbf{r}) = 0 \tag{5.25}$$

This corresponds to a steady-state version of equation (4.47) since, according to Fermi Age theory, the source of thermal neutrons is the slowing-down density of neutrons reaching thermal energy, $q_{E_{th}}$.

The slowing-down density of neutrons at fission energy is simply the total number of neutrons produced by fission:

$$q_{E_f}(\mathbf{r}) = \epsilon\nu\bar{\Sigma}_f\phi_{th}(\mathbf{r}),$$

and consequently:

$$q_{E_{th}}(\mathbf{r}) = \epsilon p\nu \exp\left(-B^2\tau_{th}\right)\bar{\Sigma}_f\phi_{th}(\mathbf{r})$$
$$= k_\infty \exp\left(-B^2\tau_{th}\right)\bar{\Sigma}_a\phi_{th}(\mathbf{r}) \tag{5.26}$$

since:

$$k_\infty\bar{\Sigma}_a = \epsilon p\nu\bar{\Sigma}_f$$

If equation (5.26) for $q_{E_{th}}(\mathbf{r})$ is substituted into equation (5.25), $\nabla^2\phi_{th}(\mathbf{r})$ is replaced by $-B^2\phi_{th}(\mathbf{r})$, and the resulting equation is divided throughout by $\bar{\Sigma}_a\phi_{th}(\mathbf{r})$, we get:

$$B^2L_c^2 + 1 = k_\infty \exp\left(-B^2\tau_{th}\right)$$

or

$$k_\infty = (1 + B^2L_c^2)\exp\left(B^2\tau_{th}\right) \tag{5.27}$$

This is the Fermi Age critical equation, and may be compared with the two-group and modified one-group critical equations, (5.10) and (4.61). In particular it should be noted that if the size of the reactor is large, B^2 is very small and $\exp\left(B^2\tau_{th}\right) \approx 1 + B^2\tau_{th}$. In this case the Fermi Age equation becomes:

$$k_\infty = (1 + B^2L_c^2)(1 + B^2\tau_{th})$$

which is identical with the two-group equation, and very nearly the same as the modified one-group equation. All these equations are similar in that they relate the materials of a reactor, as expressed by k_∞, L_c^2 and L_s^2 (or τ_{th}), to the dimensions of the reactor as expressed by B^2, whose values for different core shapes are given in Table 4.3 (p. 102).

Equation (5.27) is transcendental in B^2, and if B^2 is unknown it must be found by an iterative procedure. A good first approximation for B^2 is the value given by modified one-group theory, namely:

$$\frac{k_\infty - 1}{L_c^2 + L_s^2}$$

The leakage probabilities for fast and thermal neutrons are given by Fermi Age theory as:

$$P_{Lf} = 1 - \exp\left(-B^2\tau_{th}\right) \qquad (5.28)$$

and

$$P_{Lth} = \frac{B^2 L_c^2}{1 + B^2 L_c^2}$$

as in one-group theory.

5.3　Heterogeneous reactors

There are several reasons why a heterogeneous system of fuel and moderator is preferable to a homogeneous one. In a power reactor with solid fuel (as distinct from a liquid or slurry fuel) the rate of energy release by fission is high and it is essential to circulate the coolant in close contact with the fuel. It is also important from time to time during the operating life of a reactor to be able to remove fuel from the core as it is burned up, and replace it with fresh fuel. Both these requirements dictate that the fuel should be in the form of individual elements, or bundles of elements, arranged in a regular lattice within the moderator.

There are also good reasons from the reactor physics point of view for preferring a heterogeneous reactor, as was realized in the early years of the development of the first nuclear reactors in the United States of America. At that time natural uranium was the only available nuclear fuel, and water and graphite were the only available moderators. As was pointed out in the previous chapter a critical system, even of infinite size, is not possible with natural uranium and either graphite or water. It was realized, however, that the resonance capture of neutrons in ^{238}U can be greatly reduced if the fuel is concentrated in lumps or elements which are dispersed throughout the moderator. In this way some neutrons may slow down in the moderator without ever entering the fuel, and those neutrons which do enter the fuel elements at energies corresponding to the ^{238}U resonances are captured near the surface of the fuel with the result that the ^{238}U in the interior of a fuel element is shielded from these neutrons. This effect results in a considerable increase in the value of the resonance escape probability as compared with the value for the equivalent homogeneous system with the same fuel–moderator ratio, and makes possible a critical system with natural uranium and graphite.

In many ways the analysis of a heterogeneous reactor is similar to that of a homogeneous reactor. The reproduction constant is defined in the same way, and the four factors of k_∞ have essentially the same meaning in both types of reactor; however, in a heterogeneous reactor these factors cannot be calculated by the expressions developed in the last chapter for homogeneous reactors. Once the four factors and the value of k_∞ are known, the determination of the critical size follows the method of the last chapter, using equation (4.61) and appropriate values of L_s^2 and L_c^2, which may also have to take account of the heterogeneous structure of the core.

5.4   Eta

The factor η depends on the composition of the fuel and not the moderator, so it might be expected that its value in a heterogeneous system is the same as in a homogeneous system with the same fuel. This is not strictly true because of hardening of the neutron spectrum in the fuel of a heterogeneous reactor and the non-$(1/v)$ variation of the fuel cross-section. In practice the value of η used in a heterogeneous reactor calculation may be an empirical value chosen to make calculated values of ϵ, p and f agree with a value of k_∞ determined from an exponential experiment. The empirical value can be used in further calculations in which the fuel and moderator are the same, but the lattice dimensions are altered.

5.5   Thermal utilization factor

The thermal utilization factor is affected by a change from a homogeneous to a heterogeneous system. This change can be explained by noting that thermal neutrons are produced in the moderator as a result of slowing down, but are nearly all absorbed in the fuel due to its much greater absorption cross-section. There is therefore a net flow or current of thermal neutrons from the moderator into the fuel. Since, as was shown in the last chapter, the direction of neutron current is also the direction of decreasing neutron flux, it follows that the average thermal neutron flux in the fuel is less than in the moderator.

The variation of the flux in the fuel and moderator can be determined approximately by applying one-group diffusion theory to an equivalent lattice cell consisting of a fuel element and its associated moderator, see Figure 5.1(a). If the number of fuel elements in the core is large, a single cell can be regarded as representative of the core as a whole. A typical flux variation is shown in Figure 5.1(b), which shows the depression of the flux from the moderator into the fuel, and also a slight depression in the fuel itself.

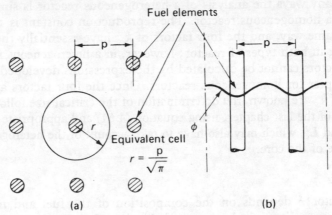

Figure 5.1. Thermal neutron flux variation in a heterogeneous reactor

The expression for the thermal utilization factor of a heterogeneous reactor is:

$$f = \frac{V_F \Sigma_{aF} \bar{\phi}_F}{V_F \Sigma_{aF} \bar{\phi}_F + V_M \Sigma_{aM} \bar{\phi}_M} \tag{5.29}$$

where V_F and V_M are the volumes of fuel and moderator, Σ_{aF} and Σ_{aM} are their macroscopic cross-sections, and $\bar{\phi}_F$ and $\bar{\phi}_M$ are the average thermal neutron fluxes in the fuel and moderator. An alternative expression for f is:

$$f = \frac{V_F \Sigma_{aF}}{V_F \Sigma_{aF} + V_M \Sigma_{aM}(\bar{\phi}_M/\bar{\phi}_F)} \tag{5.30}$$

The ratio $\bar{\phi}_M/\bar{\phi}_F$ is known as the disadvantage factor and is greater than 1, consequently the value of f is less than the value for the equivalent homogeneous system in which $\bar{\phi}_F = \bar{\phi}_M$. The decrease is about 5 per cent in the case of a natural uranium, graphite-moderated reactor with 2·5 cm diameter fuel elements in a 20 cm square lattice.

In a large reactor the fine structure flux illustrated in Figure 5.1(b) is the same in all lattice cells, and is superimposed on the overall flux shape in the core which was derived in the last chapter for the principal core geometries.

5.6 Resonance escape probability

The increase in the resonance escape probability referred to earlier in this chapter is due to two effects. In the first place neutrons whose energy corresponds to one of the resonances of ^{238}U and which diffuse from the moderator into the fuel are captured very close to the surface of the fuel. For example, neutrons of energy 6·7 eV (corresponding to the lowest energy resonance of ^{238}U whose value is about 7000 barns) have a mean free path of about 0·003 cm in natural uranium, so it is obvious that very few of these neutrons penetrate more than a fraction of a millimetre into the fuel. This is an extreme case, however the important point is that resonance energy neutrons are captured near the surface of the fuel, and the interior of the fuel is shielded from these neutrons. Stated in another way, it may be said that the flux of resonance neutrons is very much depressed in the interior of fuel elements, with a consequent reduction in the capture of these neutrons.

The production of neutrons by fission takes place throughout the fuel, consequently the fraction of these that are captured in ^{238}U resonances can be minimized by reducing the surface to volume (or mass) ratio of the fuel. For cylindrical fuel elements the surface to volume ratio is inversely proportional to the radius, so that increasing the fuel element size for a given moderator–fuel ratio has the effect of increasing the resonance escape probability.

The calculation of the resonance escape probability for a heterogeneous core uses the equivalent cell concept introduced in the last section, and an effective resonance integral which includes a surface to mass term to take account of the surface capture effect. One form of this effective resonance integral for natural uranium rods is:

$$I \,(\text{nat U}) = 9{\cdot}25 + 24{\cdot}7 \,\frac{S}{M} \,\text{barns} \tag{5.31}$$

where the surface to mass ratio, S/M, is in cm²/g.

The resonance escape probability is given by an equation of the form:

$$p = \exp - \left[\frac{N_F V_F I}{\xi_M V_M \Sigma_{sM}} \right] \tag{5.32}$$

which is similar to the corresponding expression for p in a homogeneous core.

The second effect of a homogeneous system on resonance capture is purely geometrical. If the fuel elements are widely spaced in the moderator, some neutrons may slow down in the moderator without entering the fuel and being liable to resonance capture. This effect is not as important as the one already described.

The overall result of changing from a homogeneous to a heterogeneous system of the same composition is to increase the resonance escape probability by possibly as much as 50 per cent. It is this effect which makes possible a critical system with natural uranium and graphite.

5.7   Fast fission factor

A heterogeneous arrangement also has an effect on the fast fission factor. In a homogeneous system a fission neutron emitted from the point of fission travels through a medium consisting almost entirely of moderator, with fuel atoms thinly dispersed in it, and the probability of ^{238}U fission in such a medium before the neutron's energy falls below 1 MeV is negligibly small. In a heterogeneous system, however, a fission neutron travels for a short distance through pure fuel before it passes into the moderator. During its passage through the fuel the neutron may interact, the probability depending on the neutron's track length in the fuel which in turn depends on the size of the fuel element. If an interaction does occur in a natural uranium fuel element there is (recalling the cross-sections quoted in Chapter 3 for 2 MeV neutrons in ^{238}U) about a 1 in 6 chance that it is fission in ^{238}U. Consequently the probability of fast fission is slightly greater in a heterogeneous reactor than in a homogeneous reactor, the magnitude of the effect depending on the size of the fuel elements. For 2·5 cm diameter natural uranium fuel elements the value of the fast fission factor is about 1·03.

5.8   Diffusion length and slowing-down length

If neutron absorption in the cladding, coolant and structural materials is negligible, and if the volume of fuel is small compared with the volume of moderator, the diffusion length squared for the core may be determined by the equation derived in the last chapter:

$$L_c^2 = (1 - f)L_M^2$$

where L_M is the diffusion length of pure moderator. This expression may have to be modified if gas-filled coolant channels (which are for all practical purposes voids as far as neutrons are concerned) permit neutron streaming and increased leakage in the direction of the channels.

The slowing-down length squared (or age) may also have to be modified in heterogeneous assemblies. Although heavy fuel atoms are ineffective at slowing down neutrons by elastic scattering, they are very effective at slowing down high energy neutrons by inelastic scattering. The age of fuel–moderator assemblies can be calculated by methods which take into account both elastic and inelastic scattering, but these methods are beyond the scope of this book. The results of such calculations for uranium–water mixtures, however, have shown that the age of the mixture is nearly the same as the age of pure water, presumably because inelastic scattering of high energy neutrons by the uranium is nearly as effective as elastic scattering by water.

In conclusion it must be emphasized that the approximate nature of some of the theory of heterogeneous reactors makes it necessary to obtain further data from experiments such as the exponential and "approach to critical" experiments. In this way calculated results can be verified, and empirical quantities such as η can be determined.

5.9 Reflected reactors

Hitherto we have considered bare reactors, which implies that neutrons leaking from the core cannot be scattered back into it. A bare core is impracticable in a power reactor as there must be some structural material and shielding surrounding it, and from the point of view of neutron economy it is clearly desirable that some neutrons leaking from the core should be scattered and return to it. In this way neutron leakage is reduced, and the critical size and mass of fuel can also be reduced.

It is usual, therefore, for a reactor to be built with its core surrounded by a reflector. The primary function of the reflector is to scatter neutrons back into the core, it should therefore be a material of high scattering cross-section and low capture cross-section. Furthermore, if the reflector is an element (or contains an element) of low mass number, then fast neutrons leaking into the reflector may not only be scattered back into the core, but slowed down in the process, which is an advantage in a thermal reactor. From these points it is obvious that a good moderator is also a good reflector, and in many thermal reactors the moderator and the reflector are of the same material.

A further advantage is that the thermal neutron flux in the core of a reflected reactor is more uniform than in a bare core, and consequently the flux form factors are closer to unity. This has the advantages in a power reactor that the rate of burnup of the fuel and the outlet temperature of the coolant are more nearly uniform across the core.

5.10 One-group theory of reflected reactors

The simplest theory for the analysis of reflected reactors is one-group theory. The method used for bare reactor analysis by separation of variables is limited because of the difficulty of satisfying the boundary conditions at the core–reflector interface. (This is because the reflector is more effective at the corners of the core than at other points on its surface.) It is possible to analyse a reflected spherical reactor as this involves only one coordinate, and the results can be applied to cylindrical and rectangular cores.

Consider a spherical reactor core of radius R_c, surrounded by a reflector of outer radius R_R (extrapolated radius R^+), see Figure 5.2(a). The core of the reactor has material properties defined by its infinite reproduction constant k_∞, diffusion coefficient D_c and diffusion length L_c. The diffusion coefficient and diffusion length of the reflector are D_R and L_R. Following the method used in the last chapter for the analysis of bare reactors, the one-group diffusion equation for the core is written using spherical coordinates and assuming angular symmetry as:

$$\frac{d^2\phi_c}{dr^2} + \frac{2}{r}\frac{d\phi_c}{dr} + B_c^2\phi_c = 0 \qquad (5.33)$$

where:

$$B_c^2 = \frac{k_\infty - 1}{L_c^2}$$

The one-group equation for the reflector is:

$$\frac{d^2\phi_R}{dr^2} + \frac{2}{r}\frac{d\phi_R}{dr} - \kappa_R^2\phi_R = 0 \qquad (5.34)$$

where:

$$\kappa_R = \frac{1}{L_R}$$

The solution of equation (5.33) is:

$$\phi_c = \frac{A \sin B_c r}{r} + \frac{C \cos B_c r}{r}$$

and the constant $C = 0$, otherwise ϕ_c would become infinite as $r \to 0$, therefore:

$$\phi_c = \frac{A \sin B_c r}{r} \qquad (5.35)$$

The solution of equation (5.34) is:

$$\phi_R = \frac{A' \sinh \kappa_R r}{r} + \frac{C' \cosh \kappa_R r}{r} \qquad (5.36)$$

To satisfy the boundary condition at the outer surface of the reflector, namely $\phi_R(R^+) = 0$, we get:

$$C' = -A' \tanh \kappa_R R^+ \qquad (5.37)$$

The conditions of continuity of neutron flux and current at the interface between the core and the reflector yield the equations:

$$A \sin B_c R_c = A'(\sinh \kappa_R R_c - \tanh \kappa_R R^+ \cosh \kappa_R R_c) \quad (5.38)$$

and

$$D_c A(B_c R_c \cos B_c R_c - \sin B_c R_c)$$
$$= D_R A'(\kappa_R R_c \cosh \kappa_R R_c - \sinh \kappa_R R_c$$
$$- \tanh \kappa_R R^+ \{\kappa_R R_c \sinh \kappa_R R_c - \cosh \kappa_R R_c\}) \quad (5.39)$$

Dividing equation (5.39) by (5.38) we get:

$$D_c(B_c R_c \cot B_c R_c - 1) = -D_R(\kappa_R R_c \coth \kappa_R \{R^+ - R_c\} + 1)$$

or, if the extrapolation distance is neglected (implying that $R^+ = R_R$):

$$D_c(B_c R_c \cot B_c R_c - 1) = -D_R(\kappa_R R_c \coth \kappa_R T + 1) \quad (5.40)$$

where T is the thickness of the reflector.

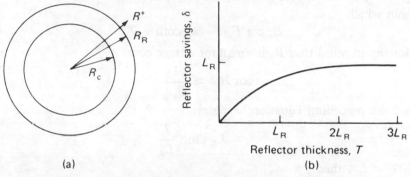

Figure 5.2. Reflected spherical reactor and reflector savings

Equation (5.40) gives the relationship between the buckling, core size and reflector thickness for a spherical reactor, and may be compared with the corresponding equation for the bare spherical core which is $B = \pi/R'$, where R' is the extrapolated radius of the bare core. Equation (5.40) is an eigenvalue equation, and it is the fundamental solution of this equation, that is the one giving the lowest value of $B_c R_c$, that determines the critical condition. The equation may be used to determine the core size of a reflected reactor whose composition (B_c, D_c, κ_R, D_R) and reflector thickness (T) are known, in which case the equation is solved for R_c. Alternatively if the unknown quantity is the composition of the core and the core size and reflector thickness are known, equation (5.40) is solved for B_c and the composition determined using equation (4.56).

While the one-group method is not particularly accurate in some respects, it does predict quite accurately the reduction in the core size that results from using a reflector. The term reflector savings is used to denote this reduction in core size. For a spherical core

Reflector savings = Radius of bare core
                    − Radius of reflected core of the same composition.

Neglecting the extrapolation distance, the radius of a bare spherical core of buckling B_c is equal to π/B_c.

The reflector savings is:

$$\delta = \frac{\pi}{B_c} - R_c$$

therefore:

$$B_c R_c = \pi - B_c \delta$$

Substituting this expression for $B_c R_c$ into equation (5.40), and making the simplifying assumption that the moderator and the reflector are the same material, which is quite usual in practice and means that $D_c = D_R$, we get:

$$B_c \cot(\pi - B_c \delta) = -\kappa_R \coth \kappa_R T$$

from which

$$B_c \cot B_c \delta = \kappa_R \coth \kappa_R T$$

Bearing in mind that B_c is small for a large core:

$$\cot B_c \delta \approx \frac{1}{B_c \delta}$$

and the preceding equation becomes:

$$\delta \approx L_R \tanh \frac{T}{L_R}$$

If $T < L_R$, then:

$$\tanh \frac{T}{L_R} \approx \frac{T}{L_R} \quad \text{and} \quad \delta \approx T$$

In other words if the reflector thickness is small compared with the diffusion length of the reflector, then the reflector savings is approximately equal to the reflector thickness.

On the other hand, if $T > L_R$:

$$\tanh \frac{T}{L_R} \to 1 \quad \text{and} \quad \delta \to L_R$$

This implies that as the reflector thickness increases, the reflector savings reaches a limiting value equal to the diffusion length of the reflector. In fact this limiting value is nearly reached when the reflector thickness is twice the diffusion length since tanh 2 = 0·964, so there is no point in making the reflector thickness greater than about twice the diffusion length of the reflector material. Figure 5.2(*b*) shows the variation of reflector savings with reflector thickness.

This conclusion regarding the reflector savings for a spherical reactor can be applied to other reactor shapes provided the size of the reactor is large enough to make $B_c\delta$ small. Furthermore the result is quite accurate and agrees well with more rigorous calculations. The usefulness of the reflector savings concept is that it can be used with bare reactor calculations to determine the size of a reflected reactor. The bare reactor calculations may be performed using modified one-group theory (which will give a more accurate result than one-group theory), and the size of the reflected reactor may be obtained by subtracting from the bare core size the reflector savings corresponding to the chosen reflector thickness.

It can be seen by comparing the flux equation for a reflected spherical core, (5.35), with the corresponding equation for a bare spherical core in Table 4.3 (p. 102) that the flux shape is the same in both provided their values of B^2 are the same. This applies also to other core shapes. The size of the reflected core is less than the bare core, and the extrapolation distance for the reflected core is effectively increased by an amount equal to the reflector savings.

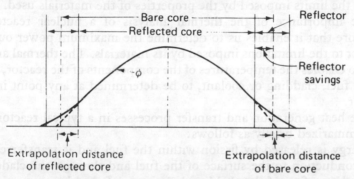

Figure 5.3. Comparison between one-group fluxes in bare and reflected cores of equal buckling

The comparison between the fluxes in a bare and a reflected core is shown in Figure 5.3, in which the solid line shows the flux variation in the core and the reflector, and the dotted line shows the variation of the flux near the edge of the equivalent bare core. It is evident that the flux in the reflected core is more nearly uniform, and that as a result the form factor is more nearly equal to one than in the equivalent bare core. The engineering advantages of this will become evident in the next chapter.

Chapter 6

Heat transfer and fluid flow in nuclear reactors

From the physical point of view there is no limit to the rate at which fission can take place, as is evident from the explosive rate of fission and energy release in an atomic bomb. From the practical point of view, however, the rate of fission in a nuclear reactor is limited by the fact that the energy released must be transferred as heat from the fuel to the coolant while maintaining the temperatures in all parts of the reactor below the limits imposed by the properties of the materials used.

The importance of the thermal analysis of a nuclear reactor is therefore that it enables us to determine the maximum power output subject to the limitations imposed by its materials. The thermal analysis also enables the temperatures of the components of the reactor, such as the fuel, cladding or coolant, to be determined at any point in the reactor.

The heat generation and transfer processes in a typical reactor can be summarized briefly as follows:

Energy is released by fission within the fuel and is transferred by heat conduction to the surface of the fuel and through the cladding. From the surface of the cladding heat is transferred by convection to the coolant. In order to promote effective heat transfer (a high heat transfer coefficient), the coolant is circulated at high velocity across the surface of the cladding so that heat transfer is by forced convection to a turbulently flowing fluid. Finally, the energy released by fission, having been transferred to the coolant, is transported out of the reactor as the coolant passes from the core to external heat exchangers in which steam may be generated for a thermodynamic power system.

In this chapter we will study these processes to determine temperatures, heat transfer rates and power levels in a reactor. The flow of coolant in a reactor will also be studied to determine friction effects, pressure losses and pumping power requirements. Only reactors using single-phase coolants, either liquids or gases, will be studied in any detail, and two-phase coolants, boiling heat transfer and the important class of boiling water reactors will only be mentioned briefly.

It will be assumed that the reader is familiar with the basic principles and equations of heat transfer and fluid flow, and these will be used as the starting points for deriving equations that are peculiar to nuclear reactors.

6.1 Heat conduction in fuel elements

The fuel elements are usually long cylindrical rods or rectangular plates of uranium or other fissionable material enclosed by cladding. The uranium may be in the pure metallic form, in the form of a compound such as uranium oxide, UO_2, or in the form of an alloy with another metal such as aluminium or zirconium. The desirable properties of the fuel, in addition to the all-important requirement that it should be fissionable, are high thermal conductivity, good corrosion resistance, good mechanical strength at high temperatures and a high limiting temperature for operation. The maximum permissible temperature of the fuel is one of the most important factors in the thermal design of a reactor.

The cladding serves three functions:

1. To prevent the release of radioactive fission products into the coolant stream.
2. To provide structural support and strength for the fuel and prevent distortion.
3. In the case of certain types of reactors (mainly gas-cooled reactors), to provide extended surfaces in the form of fins to promote more effective heat transfer to the coolant.

Suitable materials for the cladding should satisfy a number of requirements of which the most important are low neutron capture cross-section, high thermal conductivity, good mechanical strength at high temperatures and chemical compatibility with the fuel and coolant. The most common cladding materials are aluminium, magnesium alloys (Magnox), stainless steel and alloys of zirconium (Zircaloy).

The basic equation for one-dimensional heat conduction is:

$$Q = -kA \frac{dT}{dx} \qquad (6.1)$$

where Q is the rate of heat transfer, watts; A is the area through which heat is transferred, m^2; dT/dx is the temperature gradient at the point considered, K/m; and k is the thermal conductivity of the material, W/m K. (The negative sign indicates that the heat flow is in the direction of decreasing temperature.)

Alternatively equation (6.1) may be written as:

$$\frac{Q}{A} = q = -k \frac{dT}{dx} \qquad (6.2)$$

where q is the heat flux, W/m^2. The similarity between equation (6.2) for heat flux and equation (4.18) for neutron current should be noted.

The heat conduction equation may be generalized to a three-dimensional equation by the same method as that used in Chapter 4 to express the rate of neutron leakage per unit volume. The result is:

The rate of heat loss per unit volume by conduction:

$$= -k\nabla^2 T \tag{6.3}$$

The similarity between this equation and equation (4.24) for neutron leakage is evident, and the Laplacian operator ∇^2 can be expanded in rectangular, cylindrical or spherical coordinates according to the nature of the problem.

The general heat conduction equation for a medium in which energy is being released (possibly by fission) is, per unit volume:

$$
\begin{pmatrix} \text{The rate of change of the} \\ \text{internal energy of the medium} \end{pmatrix} = \begin{pmatrix} \text{The rate of energy} \\ \text{release in the medium} \end{pmatrix}
$$
$$
- \begin{pmatrix} \text{The rate of heat conduction} \\ \text{out of the medium} \end{pmatrix}
$$

$$\rho c \frac{\partial T}{\partial t} = H + k\nabla^2 T \tag{6.4}$$

where ρ is the density of the material, kg/m^3; c is the specific heat of the material, $J/kg\ K$; and H is the rate of energy release in the medium, W/m^3.

We are concerned in this chapter with steady-state conditions in a reactor, so temperatures are independent of time. Equation (6.4) simplifies in this case to:

$$\nabla^2 T = -\frac{H}{k} \tag{6.5}$$

and this equation will be used to analyse the temperatures in reactor fuel elements in which energy is being released by fission.

If it is assumed that all the energy released by fission appears in the fuel, then the rate of energy release per unit volume of fuel is given by:

$$H = 3\cdot2 \times 10^{-5}\ \Sigma_f \phi_{th}\ W/m^3$$

where Σ_f is the average macroscopic fission cross-section of the fuel. Alternatively the rate of energy release may be expressed per unit mass of fuel as the rating R:

$$R = \frac{3\cdot2 \times 10^{-8}\ \Sigma_f \phi_{th}}{\rho_F}\ W/gram\ or\ MW/tonne$$

where ρ_F is the density of the fuel, kg/m^3.

As an indication of the magnitude of these energy ratings, it may be calculated that in a reactor fuelled with enriched uranium dioxide fuel of density 10 000 kg/m³, with 3 per cent ^{235}U in the uranium, and in a thermal neutron flux of 7×10^{13} neutrons/cm²s (which is approximately the value of the maximum thermal neutron flux in a pressurized water reactor), the energy release per unit volume of fuel is about 800 MW/m³ of UO_2, and the energy release per unit mass of fuel is about 80 MW/ tonne of UO_2. These figures may be regarded as typical of the maximum heat ratings in PWRs.

The total thermal output of a reactor Q_r is related to the maximum volumetric energy release rate H_{max} and the maximum rating R_{max} by the equations:

$$Q_r = \frac{H_{max} \times \text{Volume of fuel}}{\alpha_z \times \alpha_r}$$

and

$$Q_r = \frac{R_{max} \times \text{Mass of fuel}}{\alpha_z \times \alpha_r}$$

where α_z and α_r are the axial and radial form factors respectively.

In the thermal analysis of a reactor we are usually interested in the maximum temperatures in the core which normally occur in the most highly rated fuel element and its associated coolant. If the fuel distribution in the core is uniform, the most highly rated element is the central one, and we will consider it. The flux variation along the central fuel element is given by:

$$\phi(z) = \phi_{max} \cos \beta z$$

where $\beta = \pi/L'$, and the rate of energy release is given by:

$$H(z) = H_{max} \cos \beta z$$

It may be assumed that since we are usually considering thin cylindrical or plate type fuel elements the flux, and hence the energy release rate, are uniform across the fuel. This assumption thus neglects the flux depression within the fuel that was described in the previous chapter, however this depression is small if the fuel thickness is small.

6.2 Plate type fuel element

Figure 6.1 shows a plate type fuel element in which the fuel of thickness $2a$ is enclosed on both sides by cladding of thickness b. The dimensions of the plate in the y and z directions are large compared with the values of a and b, consequently heat conduction may be assumed to be in the direction of the x-axis only.

In the fuel, equation (6.5) is written as:

$$\frac{d^2T}{dx^2} = -\frac{H}{k_F}$$

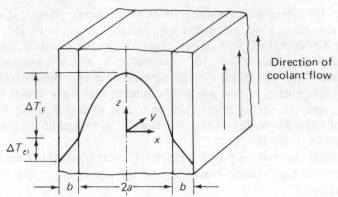

Figure 6.1. Temperature distribution in a plate type fuel element

Assuming constant values for the thermal conductivity and the energy release rate, this equation when integrated becomes:

$$\frac{dT}{dx} = -\frac{Hx}{k_F} + C$$

When $x = 0$, $dT/dx = 0$, therefore $C = 0$. Integrating again between $x = 0$ and $x = a$, we get:

$$\Delta T_F = \frac{Ha^2}{2k_F} \tag{6.6}$$

where ΔT_F is the temperature drop from the centre to the surface of the fuel. The temperature distribution within the fuel is parabolic.

In the cladding there is no energy release, and the heat conducted per unit area through the cladding on each side of the fuel is Ha watts/m². Using equation (6.2):

$$\frac{dT}{dx} = -\frac{Ha}{k_{Cl}}$$

and integrating from $x = a$ to $x = a + b$, we get:

$$\Delta T_{Cl} = \frac{Hab}{k_{Cl}} \tag{6.7}$$

where ΔT_{Cl} is the temperature drop through the cladding. The total temperature drop from the centre of the fuel to the surface of the cladding (neglecting any temperature drop across the fuel–cladding interface), is:

$$\Delta T_F + \Delta T_{Cl} = Ha\left(\frac{a}{2k_F} + \frac{b}{k_{Cl}}\right) \tag{6.8}$$

6.3 Cylindrical fuel element

Figure 6.2 shows a cylindrical fuel rod of radius a surrounded by

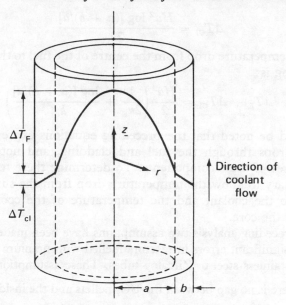

Figure 6.2. Temperature distribution in a cylindrical fuel element

cladding of thickness b. If heat conduction along the rod is negligible, which is the case if the length of the rod is much greater than its radius, then the heat conduction equation for the fuel in cylindrical coordinates is (referring back to the last lines of Section 4.10):

$$\frac{1}{r} \frac{d}{dr} \left(r \frac{dT}{dr} \right) = -\frac{H}{k_F} \qquad (6.9)$$

or:

$$\frac{d}{dr} \left(r \frac{dT}{dr} \right) = -\frac{Hr}{k_F}$$

Integrating, we get:

$$\frac{dT}{dr} = -\frac{Hr}{2k_F} + \frac{c}{r}$$

When $r = 0$, $dT/dr = 0$, therefore $c = 0$.

Integrating again between $r = 0$ and $r = a$, the temperature drop from the centre to the surface of the fuel is given by:

$$\Delta T_F = \frac{Ha^2}{4k_F} \qquad (6.10)$$

The heat conducted through the cladding per unit length of fuel element $= \pi a^2 H$, and using the well-known equation for the temperature drop through a cylindrical wall, the equation for the temperature drop through the cladding is:

$$\Delta T_{Cl} = \frac{Ha^2 \log [(a + b)/a]}{2k_{Cl}} \qquad (6.11)$$

The total temperature drop from the centre of the fuel to the surface of the cladding is:

$$\Delta T_F + \Delta T_{Cl} = \frac{Ha^2}{2} \left(\frac{1}{2k_F} + \frac{\log [(a + b)/a]}{k_{Cl}} \right) \qquad (6.12)$$

It should be noted that the preceding equations express the temperature drops through the fuel and cladding, and not the actual temperatures of these components. To determine these temperatures it is necessary to know the temperature drop from the surface of the cladding to the coolant and the temperature of the coolant at any position in the core.

In the preceding analysis two assumptions have been made which may introduce significant errors in reactors fuelled with uranium dioxide fuel pellets in stainless steel or Zircaloy tubes. These assumptions are:

1. That there is no gap between the UO_2 pellets and the inside of the fuel tube.
2. That the thermal conductivity of the UO_2 fuel is independent of its temperature.

In the manufacture of UO_2 fuel rods a radial gap of about $0 \cdot 1$ mm is left between the UO_2 pellets and the fuel tube. At the reactor's operating temperature, due to thermal expansion of the fuel and as a result of swelling and cracking of the fuel after long irradiation, the gap decreases to about $0 \cdot 01$ mm and may close altogether in a non-uniform manner. Initially the gap is filled with helium, but as irradiation of the fuel proceeds the fission product gases, krypton and xenon, are released from the fuel and diffuse through cracks into the gap, altering its heat transfer characteristics.

Heat transfer across the gap may be characterized by the gap conductance, h_G W/m^{2}K, where:

$$h_G = \frac{k}{g}$$

where k is the thermal conductivity of the gas in the gap, W/mK and g is the width of the gap, m.

As an example, the thermal conductivity of helium at 600°C is $0 \cdot 33$ W/mK and if the gap is taken to be $0 \cdot 1$ mm, then the gap conductance is equal to 3300 W/m^{2}K. This value tends to increase as the fuel is irradiated and the gap closes.

The temperature drop across the gas gap for a cylindrical fuel rod is given by:

$$\Delta T_G = \frac{Ha}{2h_G} \qquad (6.13)$$

and equation (6.12) for the total temperature drop from the centre of the fuel to the surface of the cladding may be modified to take account of this gap as:

$$\Delta T_{\text{total}} = \Delta T_{\text{F}} + \Delta T_{\text{G}} + \Delta T_{\text{cl}} =$$
$$\frac{Ha^2}{2}\left(\frac{1}{2k_{\text{F}}} + \frac{1}{ah_{\text{G}}} + \frac{\log[(a+b)/a]}{k_{\text{cl}}}\right) \qquad (6.14)$$

The second assumption concerns the thermal conductivity of the UO_2 fuel, which has been assumed in equation (6.6) etc. to be constant. The thermal conductivity of UO_2 is much lower than that of pure metallic uranium, and consequently the temperature drop across a UO_2 pellet is much larger than in a similar uranium fuel rod. With this large variation of temperature within the fuel, the assumption of a constant thermal conductivity may introduce a significant error.

If the thermal conductivity of the fuel is assumed to be a function of its temperature, the general heat conduction equation for cylindrical coordinates, equation (6.9), must be modified to:

$$\frac{1}{r}\frac{d}{dr}\left(rk_{\text{F}}\frac{dT}{dr}\right) = -H \qquad (6.15)$$

By integration the solution of this equation can be written as:

$$\int_{T_f}^{T_F} k_{\text{F}}\,dT = \frac{Ha^2}{4} \qquad (6.16)$$

where T_F is the fuel temperature at the centre line of the pellet, and T_f is the temperature at the surface of the pellet.

Values for the thermal conductivity of UO_2 have been determined and expressed as $\int_{500}^{T} k_{\text{F}}\,dT$, where the temperature 500°C has been selected as the datum, and the integral is plotted against T in Figure 6.3. The left hand side of equation (6.16) can be expressed as:

$$\int_{T_f}^{T_F} k_{\text{F}}\,dT = \int_{500}^{T_F} k_{\text{F}}\,dT - \int_{500}^{T_f} k_{\text{F}}\,dT \qquad (6.17)$$

The use of equations (6.16) and (6.17) to determine the temperature in a UO_2 fuel pellet can be illustrated by the following example:

The rate of energy release at some point in a UO_2 fuel rod is 800 MW/m^2, the fuel pellet diameter is 8·3 mm and its surface temperature is 600°C. Determine the temperature at the centre of the pellet.

Equations (6.16) and (6.17) can be written as:

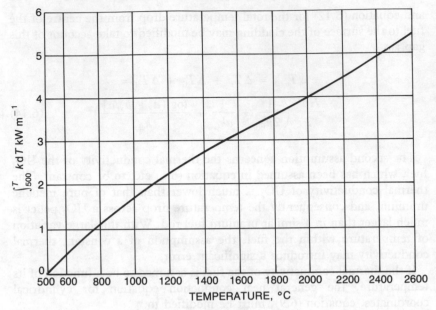

Figure 6.3. $\int_{500}^{T} K\,\mathrm{d}T$ for UO_2

$$\int_{500}^{T_F} k_F \,\mathrm{d}T = \frac{Ha^2}{4} + \int_{500}^{600} k_F \,\mathrm{d}T$$

The value of the integral on the right side of this equation is found from Figure 6.3 to be $0\cdot5$. Thus:

$$\int_{500}^{T_F} k_F \,\mathrm{d}T = 3\cdot94$$

From Figure 6.3 the value of T_F which satisfies this equation is 1980°C. This is the temperature at the centre of the fuel pellet.

6.4 Convective heat transfer from the fuel element to the coolant

The general equation for convective heat transfer between a surface and a fluid flowing across that surface is:

$$Q = hA\theta \quad \text{or} \quad q = h\theta \qquad (6.18)$$

where θ is the difference between the surface temperature and the bulk temperature of the fluid flowing across it, °C; and h is the heat transfer coefficient, W/m² °C. The very important matter of determining heat transfer coefficients will be dealt with later in this chapter.

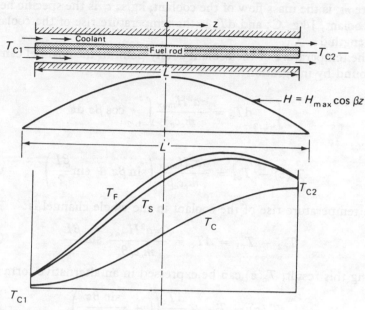

Figure 6.4. Coolant flow and temperatures in the central fuel channel

In the case of the plate type fuel element, the temperature drop at any point from the surface of the cladding to the coolant $\theta_c(z)$ is given by:

$$\theta_c(z) = \frac{H(z)a}{h} \tag{6.19}$$

and in the case of the cylindrical fuel element $\theta_c(z)$ is given by:

$$\theta_c(z) = \frac{H(z)a^2}{2(a + b)h} \tag{6.20}$$

The total temperature drop from the centre of the fuel to the coolant is $(\Delta T_F + \Delta T_{C1} + \theta_c)$.

The coolant temperature at any point in the core depends upon its inlet temperature and the heat transferred to it up to the point in question.

Figure 6.4 shows a cylindrical fuel element of length L in its coolant channel. The extrapolated length of the core is L'. The coolant inlet and outlet temperatures are T_{C1} and T_{C2}. It will be assumed as before that there is no axial heat conduction in the fuel element, in other words all the energy released in a section of the fuel is transferred radially outwards to the coolant flowing past that section. An energy balance for the coolant as it flows past a section of length dz at z is:

$$\dot{m}_c c_c \, dT_c = \pi a^2 H_{max} \cos \beta z \, dz$$

where $\dot{m}_c$ is the mass flow of the coolant, kg/s; c_c is the specific heat of the coolant, J/kg °C; and dT_c is the temperature rise of the coolant in the length dz, °C.

The temperature of the coolant at any point in the channel can now be found by integration:

$$\int_{-L/2}^{z} dT_c = \frac{\pi a^2 H_{max}}{\dot{m}_c c_c} \int_{-L/2}^{z} \cos \beta z \, dz$$

hence:

$$T_c(z) - T_{c1} = \frac{\pi a^2 H_{max}}{\dot{m}_c c_c \beta} \left(\sin \beta z + \sin \frac{\beta L}{2} \right) \tag{6.21}$$

The temperature rise of the coolant in the whole channel is:

$$T_{c2} - T_{c1} = \Delta T_c = \frac{2\pi a^2 H_{max}}{\dot{m}_c c_c \beta} \sin \frac{\beta L}{2} \tag{6.22}$$

Using this result, $T_c(z)$ can be expressed in an alternative form as:

$$T_c(z) = T_{c1} + \frac{\Delta T_c}{2} \left(1 + \frac{\sin \beta z}{\sin \beta L/2} \right) \tag{6.23}$$

The coolant temperature varies as a sine function as shown in Figure 6.4.

The temperatures at the surface of the cladding T_s, at the fuel-cladding interface T_{FC1}, and at the centre of the fuel T_F, can now be found at any position z. For a cylindrical fuel element the equations are:

$$T_s(z) = T_c(z) + \theta_c(z)$$

$$= T_{c1} + \frac{\Delta T_c}{2} \left(1 + \frac{\sin \beta z}{\sin (\beta L/2)} \right) + \theta_{c0} \cos \beta z \tag{6.24}$$

$$T_{FC1}(z) = T_c(z) + \theta_c(z) + \Delta T_{c1}(z)$$

$$= T_{c1} + \frac{\Delta T_c}{2} \left(1 + \frac{\sin \beta z}{\sin (\beta L/2)} \right)$$

$$+ \theta_{c0} \cos \beta z \left(1 + \frac{(a + b)h \log [(a + b)/a]}{k_{C1}} \right) \tag{6.25}$$

$$T_F(z) = T_c(z) + \theta_c(z) + \Delta T_{C1}(z) + \Delta T_F(z)$$

$$= T_{c1} + \frac{\Delta T_c}{2} \left(1 + \frac{\sin \beta z}{\sin (\beta L/2)} \right)$$

$$+ \theta_{co} \cos \beta z \left(1 + \frac{(a + b)h \log [(a + b)/a]}{k_{Cl}} + \frac{(a + b)h}{2k_F} \right)$$

$$(6.26)$$

where:

$$\theta_{co} = \frac{H_{max} a^2}{2(a + b)h}$$

the temperature drop from the surface of the cladding to the coolant at the mid-point of the channel. In the above equations (6.25) and (6.26) the temperature drop across the gas gap at the fuel-cladding interface is neglected.

Figure 6.4 shows the variation of T_c, T_s and T_F along the coolant channel and shows the existence of maximum values in T_s and T_F. It is important to determine the magnitude of these maximum values. Differentiating equation (6.24) and equating to zero, the position of the maximum cladding surface temperature is given by the equation:

$$z = \frac{1}{\beta} \tan^{-1} \frac{\Delta T_c/2}{\theta_{co} \sin (\beta L/2)}$$

$$(6.27)$$

Substituting this expression for z into equation (6.24) the maximum cladding surface temperature is given by:

$$T_{s(max)} = T_{c1} + \frac{\Delta T_c}{2} \left[1 + \sqrt{\operatorname{cosec}^2 \frac{\beta L}{2} + \left(\frac{\theta_{co}}{\Delta T_c/2} \right)^2} \right] \quad (6.28)$$

By the same procedure the maximum fuel temperature may be found to be given by:

$$T_{F(max)} = T_{c1} + \frac{\Delta T_c}{2} \left[1 + \sqrt{\operatorname{cosec}^2 \frac{\beta L}{2} + \left(\frac{C}{\Delta T_c/2} \right)^2} \right] \quad (6.29)$$

where:

$$C = \theta_{co} \left(1 + \frac{(a + b)h \log [(a + b)/a]}{k_{Cl}} + \frac{(a + b)h}{2k_F} \right) \quad (6.30)$$

The preceding equations enable the temperatures of the fuel, cladding and coolant to be determined in terms of the reactor power. From the thermodynamic point of view it is desirable that the coolant temperature should be as high as possible as the coolant acts as the heat source for the power cycle; however, limitations on the maximum temperature of certain components of the reactor impose a limitation on the coolant temperature and the reactor power. These limitations are due to the properties of the materials used in the reactor, and some examples are:

1. The metallurgical phase change of pure uranium at 660°C produces changes in the crystalline structure of the metal. If this temperature is exceeded repeatedly, the uranium suffers considerable dimensional change and this distortion may rupture the cladding.
2. The temperature of the cladding should not be high enough to allow its creep strength to drop seriously.
3. The temperatures of all materials in the core should not be high enough to allow any chemical reactions such as corrosion or oxidation to take place on a significant scale. In carbon dioxide-cooled reactors using Magnox cladding, for example, oxidization of the Magnox limits its maximum temperature to about 525°C.
4. In a water-cooled reactor in which boiling is to be avoided, the temperature of the water must be well below the saturation temperature corresponding to the water pressure.

The derivation of the equations for the temperatures of the components of a reactor has been based on the assumption that ideal conditions exist in which all the factors which determine the temperatures are known exactly. This is of course seldom the case in a real engineering system, and in a nuclear reactor factors such as the distortion of the neutron flux, incorrect distribution of the coolant flow in the channels or slight variations in the dimensions of the fuel or cladding can cause calculated values of the temperatures to be incorrect. To allow for these uncertainties, 'hot-channel factors' are introduced which are analogous to factors of safety in engineering. Calculated temperature drops are multiplied by appropriate hot-channel factors to determine the maximum temperatures that might exist under adverse conditions.

The thermal analysis of the central channel of a reactor determines the temperatures in that channel alone. If the coolant flow is equally distributed in all channels (bearing in mind that the inlet temperature is common to all channels and the neutron flux decreases towards the sides of the core), the outlet temperature of the coolant and the maximum fuel temperature in the outer channels is less than in the central channel. The mixed temperature of the coolant flowing from the core to the heat exchangers is also less than the outlet temperature from the central channel and this is a disadvantage from the thermodynamic point of view. The mixed coolant outlet temperature can be raised if the coolant flow in the fuel element channels is controlled by means of orifices or 'gags'. If the coolant flow in each channel is proportional to the energy release in that channel, then the coolant outlet temperature from all channels is the same, and the mixed outlet temperature is raised to this value. In this case the maximum fuel temperatures in the outer channels are still less than in the central channel, and a further improvement can be made by restricting the flow in the outer channels in such a way that the maximum fuel temperatures are the same in all channels. In this way the coolant outlet temperature in the outer channels is higher than in the centre channel, and the mixed outlet temperature is raised with a consequent improvement from the thermodynamic point of view.

A method which enables the total thermal output of a reactor to be increased without increasing the maximum fuel or coolant temperatures (assuming equal coolant flow in all channels) is to vary the enrichment of the fuel in different regions of the core. Thus, if the enrichment in the outer region of the core is greater than in the central region, the fuel rating (which is proportional to the neutron flux and the fraction of ^{235}U in the fuel) can be increased in the outer region and the thermal output can be increased without exceeding the limiting temperatures in the core.

A further assumption in the preceding analysis should be mentioned, namely that all the heat release due to fission takes place in the fuel. This is not the case in a thermal reactor in which neutrons are slowed down and gamma radiation is absorbed in the moderator. The fraction of the fission energy released as heat in the moderator is about 5 per cent of the total. This energy is transferred to the coolant (unless special cooling of the moderator is provided), and the temperature rise of the coolant is the same as if all fission energy is released as heat in the fuel. However, the cladding and fuel temperatures are lower when only about 95 per cent of the heat release takes place in the fuel, and the temperature drops through the fuel and cladding are reduced by about 5 per cent. The assumption that all heat is released in the fuel overestimates the fuel and cladding temperatures and therefore errs on the safe side when determining these temperatures.

6.5 Heat transfer by forced convection

The equation for heat transfer by convection from the surface of the cladding to the coolant is:

$$Q = hA\theta \qquad (6.31)$$

The problem arises when using this equation as to how to determine the heat transfer coefficient h. The most common method for correlating convective heat transfer data is by means of dimensionless equations in which the fluid properties, system dimensions, etc. are arranged in dimensionless groups. The constants in these equations are determined by experiment. Coolant flow in a reactor is by forced convection as it is pumped through the core, and the dimensionless groups commonly used in forced convection equations are:

$$\text{The Reynolds number } (Re) = \frac{\rho v \, d_e}{\mu}$$

$$\text{The Prandtl number } (Pr) = \frac{c_p \mu}{k}$$

$$\text{The Nusselt number } (Nu) = \frac{h d_e}{k}$$

$$\text{The Stanton number } (St) = \frac{h}{\rho v c_p} = \frac{(Nu)}{(Re)(Pr)}$$

and $$\text{The Peclet number } (Pe) = (Re)(Pr)$$

The quantities involved are the fluid viscosity, μ; the fluid thermal conductivity, k; the fluid density, ρ; the fluid specific heat, c_p (the specific heat at constant pressure in the case of gases); the fluid velocity, v; and the effective diameter of the duct in which the fluid is flowing, d_e.

The effective diameter, d_e, is defined as:

$$d_e = \frac{4 \times \text{Flow area}}{\text{Wetted perimeter}} \qquad (6.32)$$

The Reynolds number, which is a measure of the ratio of the inertia to viscous forces in a flowing fluid, characterizes the flow. In particular, the transition from laminar to turbulent flow takes place at a value of (Re) of about 2000. The Prandtl number is dependent only on the properties of the fluid and is the ratio of the kinematic viscosity to the thermal diffusivity. It determines the way in which the temperature and velocity of a fluid vary near the wall of a pipe in which it is flowing and being heated or cooled.

There are many empirical dimensionless equations available for determining heat transfer coefficients. In nearly all power reactors the coolant is circulated at high velocity through the core with values of (Re) much greater than 2000, so that the flow is turbulent. We are therefore interested in equations which are applicable to turbulent flow forced convection. The most widely used of these equations for water and gases (whose values of (Pr) vary from about 0·7 upwards) in smooth round pipes is the Dittus–Boelter equation:

$$(Nu)_b = 0 \cdot 023 (Re)_b^{0 \cdot 8} (Pr)_b^{0 \cdot 4} \qquad (6.33)$$

The fluid properties in this equation are evaluated at the bulk temperature, T_b. The bulk temperature of a fluid at any point z in a duct in which it is flowing and being heated is defined by the equation:

$$T_b(z) = T_{inlet} + \frac{1}{\dot{m}c_p} \int_{inlet}^{z} q_L \, dz$$

where q_L is the heat transferred to the fluid per unit length of duct.

In the case of some fluids, certain properties (particularly the viscosity) vary considerably with temperature, and if there is a large difference between the wall temperature and the bulk temperature of the fluid it may be necessary to take this variation into account. The Sieder–Tate equation does this by the following correlation:

$$(Nu)_b = 0.023(Re)_b^{0.8}(Pr)_b^{0.4}\left(\frac{\mu_b}{\mu_w}\right)^{0.14} \tag{6.34}$$

In this equation all the properties are evaluated at the bulk temperature except μ_w which is evaluated at the wall temperature, T_w.

The Colburn equation also takes account of property variation in the fluid film near the heated wall. The equation is:

$$(Nu)_f = 0.02(Re)_f^{0.8}(Pr)_f^{0.4} \tag{6.35}$$

and the properties are evaluated at the film temperature T_f which is defined as $T_f = \frac{1}{2}(T_w + T_b)$. The difficulty in using equations (6.34) and (6.35) is that T_w and T_f are not usually known in advance, so a trial-and-error solution may be necessary.

These equations have been obtained principally from data taken in circular pipes and can generally be relied on to predict heat transfer coefficients in circular pipes to within about 10 per cent; however, they can also be applied to non-circular ducts with reasonably good accuracy. Typical coolant duct shapes encountered in nuclear reactors include annular channels between fuel elements and moderator, bundles of cylindrical fuel rods in a large cylindrical tube and assemblies of flat fuel plates in rectangular boxes. In all cases the effective diameter must be calculated in accordance with the definition already given.

The empirical equations for liquid metals are different from those for water and gases. This arises because the mechanism of heat transfer in a liquid metal is different from that in water or gases due to the dominance of conduction over turbulent diffusion as the main heat transfer process. The Prandtl numbers for liquid metals are very low for this reason. One of the accepted correlations for liquid metals which predicts heat transfer coefficients to within about 20 per cent is:

$$(Nu) = 7.0 + 0.025(Pe)^{0.8} \tag{6.36}$$

Table 6.1. gives typical values of heat transfer coefficients for gases, water and liquid metals under reactor operating conditions.

Table 6.1. *Typical values of* (Pr) *and heat transfer coefficients for reactor coolants*

	(Pr)	h W/m² °C
Gases	0·8	50 to 500
Water	1 to 7	2000 to 20 000
Liquid metals	0·01	5000 to 50 000

It is clear from the values quoted above that the heat transfer coefficients attainable with gases are quite low when compared with those attainable with water or liquid metals. In gas-cooled reactors, therefore, it may be necessary to take steps to promote more effective heat transfer from the fuel element to the coolant by the use of finned cladding. The function of the fins is to increase the surface area for heat transfer and also promote turbulence and better mixing of the gas stream. All these effects increase the heat transfer rate, however for the fins to be really effective they should be of a material of high thermal conductivity, and the design of the fins should aim to promote circulation of gas between the heated surface and the main gas stream.

The introduction of fins results in certain disadvantages. Firstly, the power required to circulate the coolant through the core is increased as a result of the increased turbulence, and secondly, unless the cladding material has a very low neutron capture cross-section, the additional neutron absorption in the fins may be unacceptable from the physics point of view. Thus fins are frequently used with Magnox clad fuel (Σ_c for magnesium = 0·0027 cm^{-1}), but are not used to any extent for stainless steel clad fuel (Σ_c for stainless steel $\approx$ 0·26 cm^{-1}). Heat transfer can also be improved without any increase in the amount of cladding used by roughening the surface of the cladding, and the heat transfer surface area can be increased by using a larger number of smaller diameter fuel rods. The use of small diameter fuel rods is also to be preferred for UO$_2$ fuel whose thermal conductivity is low.

The fins used in the Magnox clad fuel elements of British gas-cooled reactors are of several types—longitudinal, circumferential and helical. Some complicated herringbone and polyzonal designs have been produced with the aim of promoting turbulence. Heat transfer coefficients for such surfaces cannot be predicted from the empirical equations introduced earlier for smooth surfaces, and experimental data is necessary for each design. It is sufficient to reiterate that the use of extended surfaces has the effect of increasing the heat transfer rate at the expense of increased power to circulate the coolant through the core, and increased neutron absorption in the cladding.

6.6 Boiling heat transfer

The preceding section has dealt with single phase forced convection heat transfer, and is applicable to gas-cooled reactors and water-cooled reactors in which boiling does not take place. However, in certain types of reactor, notably boiling water reactors (BWRs) boiling does take place in the core of the reactor. It is beyond the scope of this book to deal in detail with the rather complex analysis of boiling heat transfer, much of which depends on experimental data expressed by means of empirical equations. The following qualitative description is intended to give no more than an introduction to the subject.

Consider convective heat transfer from a heated surface at temperature T_s (such as the surface of a fuel rod) to water whose saturation temperature is T_{sat} flowing across the surface. So long as T_s is less than T_{sat}, boiling of the water is not possible and single phase heat transfer occurs. When T_s exceeds T_{sat} it is possible that boiling may begin, dependent on the bulk water temperature T_w and the heat flux at the hot surface. So long as T_w is significantly less than T_{sat} or the heat flux is low, boiling will not occur. This is indicated by the single phase heat transfer region 1 in Figure 6.5.

However, as the heat flux increases, resulting in an increase in T_s, it is possible for boiling of the water to begin while its bulk temperature T_w is still less than T_{sat}. This is referred to as subcooled boiling. It occurs at high heat fluxes when the film temperature of the water in contact with the heated

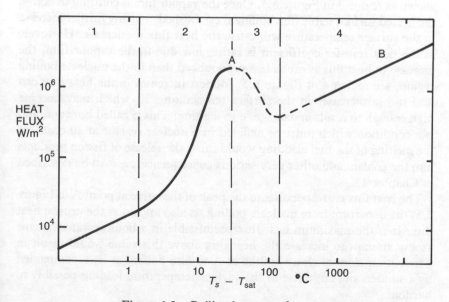

Figure 6.5. Boiling heat transfer curve

surface reaches the saturation temperature while the bulk temperature of the water is lower. As the steam bubbles formed at the heated surface move into the cooler bulk of the water they rapidly condense and collapse, but the increased turbulence caused by this bubble movement causes an increase in the heat transfer coefficient. This increasing heat transfer coefficient is indicated on Figure 6.5 by the steepening portion of the curve between regions 1 and 2 of the figure.

As the bulk water temperature rises and becomes equal to the saturation temperature, bulk boiling commences and becomes fully developed. The turbulence caused by the formation of steam bubbles and their movement away from the heated surface causes the heat transfer coefficient to increase further, resulting in the pronounced steepening of the curve in region 2 of Figure 6.5. In this region, called the nucleate boiling region, the heat flux increases rapidly with only a modest increase in the heated surface temperature.

As nucleate boiling develops, the heated surface becomes increasingly blanketed by steam bubbles which, having a thermal insulating effect, restrict heat transfer and cause a lowering of the heat transfer coefficient. The curve of Figure 6.5 begins to flatten. A heat flux is reached called the departure from nucleate boiling (DNB) at which, if the heated surface temperature is further increased, the heat transfer rate will decrease due to the increased thermal resistance of the vapour film which is spreading over the heated surface. This leads to an unstable state of partial film boiling shown as region 3 in Figure 6.5. Once the vapour film is continuous across the heated surface stable film boiling is established, and any further increase in the surface temperature will cause the heat flux to increase. However, as the heat transfer coefficient is rather low due to the vapour film, the increase in heat flux is much less pronounced than in the nucleate boiling region, see region 4 of Figure 6.5. Modest increases in the heat flux can lead to big increases in the surface temperature T_s, which may become high enough to result in damage, e.g. melting. This is called burnout, and is a condition which must be avoided in a nuclear reactor at all costs, as the melting of the fuel cladding would cause the release of fission products into the coolant and other very serious consequences, as will be described in Chapter 11.

The heat flux corresponding to the peak of the curve at point A in Figure 6.5, the departure from nucleate boiling, is also known as the critical heat flux. It is the maximum heat flux permissible in a boiling water reactor as any attempt to increase the heat flux above this value would result in a transition from nucleate boiling at A to film boiling at B, accompanied by a sudden and large rise in the surface temperature leading possibly to burnout.

It is important to be able to predict the critical heat flux in a reactor in which boiling may occur. Such reactors include not only BWRs in which nucleate boiling occurs during normal operation, but also PWRs in which boiling, either subcooled or nucleate, may occur under abnormal operating conditions. Much experimental work has been done to develop empirical correlations which involve the heat flux, surface and fluid temperatures, fluid pressure and velocity, dryness fraction (or steam quality) and the system dimensions among the relevant parameters. As a general rule it may be said that these parameters affect the critical heat flux as follows:

1. Increasing pressure may cause the critical heat flux to decrease.
2. Increasing fluid velocity causes the critical heat flux to increase.
3. Increasing dryness fraction (or steam quality) causes the critical heat flux to decrease.

As the dryness fraction increases along a channel in the reactor core in which boiling is taking place, the critical heat flux decreases along the channel, reaching a maximum at its upper end.

The heat flux during nucleate boiling can be related to the heated surface temperature by an equation of the form:

$$\frac{Q}{A} = C(T_s - T_{sat})^n \tag{6.37}$$

where C and n are constants. The value of n is typically about 4, so it is evident that the heat flux increases rapidly during nucleate boiling as the surface temperature T_s increases. This is illustrated as region 2 of Figure 6.5.

6.7 The choice of coolants for reactors

The coolant for a nuclear reactor may be either a gas, water or heavy water, or a liquid metal such as sodium or sodium–potassium alloy. In choosing the coolant for a reactor the following desirable characteristics are borne in mind:

1. Low neutron capture cross-section. This is essential as neutron economy in a reactor does not permit much neutron absorption in the coolant. It is also an advantage in a thermal reactor if the coolant has a moderating effect, i.e. low mass number and high scattering cross-section.
2. High specific heat, density, thermal conductivity and heat transfer coefficient. The first two properties determine the amount of energy per unit volume that the coolant can transport, and the other two properties control the temperature drop from the cladding to the coolant.
3. Good stability. The coolant should not react chemically with other

components of the reactor with which it is in contact, and it should be stable in conditions of high temperature and irradiation.

4. Low neutron induced radioactivity. In many power reactors the heat exchangers are unshielded and it is important that the coolant should not become radioactive in its passage through the core as a result of an (n,γ) reaction. If the coolant does become activated, the primary heat exchangers must be shielded and a secondary circuit of non-activated coolant must be used for steam generation.

Helium and carbon dioxide are the most suitable gases. The cost of helium has so far prevented its use in large power reactors in which the leakage of coolant gas may be significant; however, future high temperature gas-cooled reactors may use helium.

Water and heavy water are suitable in most respects for thermal reactors in which they can fulfil dual moderating and cooling roles. However, they have the great disadvantage that high coolant temperatures imply high system pressures and very strong containment vessels (for example, the saturation pressure corresponding to 350°C is 165 bar), and the maximum water temperature is limited to the critical temperature, 374°C. The capture cross-section of water is rather high and enriched uranium fuel is required to compensate for this; on the other hand, heavy water has a very low capture cross-section but is very expensive. Both water and heavy water require stainless steel or Zircaloy cladding.

Sodium and the eutectic sodium–potassium alloy are the most common liquid metal coolants. They have high saturation temperatures at atmospheric pressure so that a pressurized coolant system is unnecessary; however, the ^{23}Na(n,γ) ^{24}Na reaction produces a radiation hazard, the primary heat exchangers must be shielded and a secondary coolant circuit must be used for steam generation. The violent chemical reactions between sodium and air or water make it essential to eliminate the possibility of coolant leakage. The capture cross-section of sodium is rather high for its use as the coolant in a thermal reactor, and this point together with its non-moderating characteristic make it a more suitable coolant for a fast reactor.

6.8 Coolant flow through a reactor core — pressure drop and pumping power

As the coolant flows through the fuel channels in a reactor core it suffers a pressure drop which arises in two main ways, firstly friction between the coolant and the channel walls and secondly acceleration due to the decrease in coolant density as its temperature rises. The latter effect is usually small in a liquid-cooled reactor, but may be significant in a gas-cooled reactor. A third effect due to the pressure drop along the channel causing a further reduction of density and consequent acceleration is very small and will be neglected.

The friction effect in pipe flow is taken into account by the Fanning friction factor, f, which is defined by the equation:

$$\text{The shear stress at the pipe wall, } \tau_w = f \times \frac{\rho v^2}{2} \qquad (6.38)$$

The Fanning friction factor has been determined experimentally for turbulent flow in commercially smooth tubes, and the empirical correlation:

$$f = 0.046(Re)^{-0.2} \qquad (6.39)$$

is generally accepted. It is interesting to compare this equation with a modified form of the Dittus—Boelter equation, (6.33):

$$(St) = 0.023(Re)^{-0.2}(Pr)^{-0.6} \qquad (6.40)$$

If (Pr) is equal or nearly equal to unity this equation simplifies to:

$$(St) = 0.023(Re)^{-0.2} \qquad (6.41)$$

and by comparison with equation (6.39) we see that:

$$(St) = \frac{f}{2} \qquad (6.42)$$

This is in effect a statement of Reynolds analogy between heat and momentum transfer which is valid for fluids of $(Pr) \approx 1$ flowing in smooth pipes.

The coolant pressure drop in a fuel element channel can be determined by considering the channel (not necessarily of circular cross-section) shown in Figure 6.6.

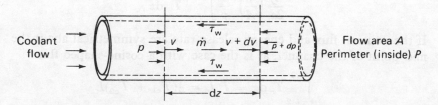

Figure 6.6. Pressure drop in a channel

The continuity equation for the flow is $\dot{m} = A\rho v$, and Newton's Second Law for the fluid in an element of the channel of length dz gives the equation:

$$-A\,dp - f\frac{\rho v^2}{2}\,P\,dz = \rho A v\,dv \qquad (6.43)$$

or, recalling that $d_e = 4A/P$:

$$-dp = \frac{4f\rho v^2}{2d_e}\,dz + \rho v\,dv \qquad (6.44)$$

Using the continuity equation to substitute $\dot{m}/A$ for ρv, and integrating along the channel from $-L/2$ to $L/2$, we get:

$$p_1 - p_2 = \frac{4}{2d_e}\left(\frac{\dot{m}}{A}\right)^2 \int_{-L/2}^{L/2} \frac{f}{\rho}\,dz + \left(\frac{\dot{m}}{A}\right)(v_2 - v_1) \qquad (6.45)$$

The friction factor f, which is proportional to $(Re)^{-0.2}$, may be affected by changes of temperature, but viscosity changes raised to the power 0.2 may be neglected and f can be regarded as constant along the length of the channel. For incompressible fluids such as water ρ and v are both constant, and the pressure drop is given by:

$$p_1 - p_2 = \frac{4fL}{2d_e\rho}\left(\frac{\dot{m}}{A}\right) = \frac{4fL\rho v^2}{2d_e} \qquad (6.46)$$

For gases, making use of the gas equation $p = R\rho T$, and assuming that pressure changes are small compared with the absolute pressure of the gas, the gas density may be expressed as a function of the temperature by the equation:

$$\frac{1}{\rho} = \frac{RT_c}{\bar{p}}$$

where $\bar{p}$ is the average gas pressure in the channel.

The first term on the right-hand side of equation (6.45) may now be written as:

$$\frac{4f}{2d_e}\left(\frac{\dot{m}}{A}\right)^2 \frac{R}{\bar{p}} \int_{-L/2}^{L/2} T_c\,dz$$

If the neutron flux and energy release rate are symmetrical about the mid-point of the channel, as is the case with a cosine-shaped flux:

$$\int_{-L/2}^{L/2} T_c\,dz = \bar{T}_c L = \tfrac{1}{2}(T_{c1} + T_{c2})L$$

The second term on the right-hand side of equation (6.45) may be written, using the continuity and gas equations, as:

$$\left(\frac{\dot{m}}{A}\right)(v_2 - v_1) = \left(\frac{\dot{m}}{A}\right)^2\left(\frac{1}{\rho_2} - \frac{1}{\rho_1}\right) = \left(\frac{\dot{m}}{A}\right)^2\frac{R}{\bar{p}}(T_{c2} - T_{c1})$$

The solution of equation (6.45) for the pressure drop in a gas is:

$$p_1 - p_2 = \left(\frac{\dot{m}}{A}\right)^2\left\{\frac{4fL}{2d_e}\frac{R\bar{T}_c}{\bar{p}} + \frac{R}{\bar{p}}(T_{c2} - T_{c1})\right\}$$

or

$$p_1 - p_2 = \frac{1}{\bar{\rho}}\left(\frac{\dot{m}}{A}\right)^2\left\{\frac{4fL}{2d_e} + \frac{T_{c2} - T_{c1}}{\bar{T}_c}\right\} \tag{6.47}$$

Typically, in a gas-cooled reactor the temperature term $(T_{c2} - T_{c1})/\bar{T}_c$ contributes about 10 per cent of the pressure drop in the core.

Equations (6.46) and (6.47) give the pressure drop in the fuel element channel only, and effects at the channel inlet and outlet, and in other parts of the coolant circuit are not included. At inlet to the core the coolant undergoes a sudden contraction in its flow as it passes from the inlet header into the fuel element channels; this is similar to the flow of fluid past a sudden contraction from a large diameter pipe into a small diameter pipe. At outlet from the core the reverse process occurs as the coolant undergoes an expansion similar to flow past a sudden enlargement in a pipe. Both these processes produce eddies and turbulence in the flow which result in losses of head, although the losses at the outlet can be reduced by using a diffuser at this end of the channel. If incompressible flow is assumed with no outlet diffuser, and the coolant velocity in the inlet and outlet headers is negligible, then the combined pressure loss at both ends of the channel is given by the approximate equation:

$$\Delta p \text{ (inlet and outlet)} = \frac{1\cdot5}{2\bar{\rho}}\left(\frac{\dot{m}}{A}\right)^2 \tag{6.48}$$

The pressure drop round the entire coolant circuit includes the pressure drop in the core plus the inlet and outlet pressure loss plus the pressure drop in the external ducts and heat exchangers. If the pressure drop round the circuit is Δp, and if this pressure drop is small compared with the absolute pressure of the coolant, then the power required to compress the coolant in the circulating pumps is given very nearly by the equation:

$$W = \frac{\dot{m}\,\Delta p}{\rho_P} \tag{6.49}$$

where ρ_P is the density of the coolant at the pumps. This power is added to the thermal output of the reactor to obtain the total energy transferred as heat to the steam. In order that W should be a minimum,

ρ_P should be as large as possible and the circulating pumps should be placed at the coolest point of the circuit, namely at inlet to the reactor. In this case $\rho_\text{P} = \rho_1$. It is also desirable from the mechanical point of view to place the pumps at the coolest point of the circuit.

Although equation (6.49) gives the power necessary to circulate the coolant, the power required to drive the pumps is greater than this by a factor $1/\eta_\text{P}$, where η_P is the combined mechanical and electrical efficiency of the pump motors. The pumping power, therefore, is given by:

$$PP = \frac{1}{\eta_\text{P}} \frac{\dot{m} \, \Delta p}{\rho_1} \tag{6.50}$$

This power must be provided by the power plant, usually as a fraction of the electrical output of the turbo-generators. The effect of the pumping power on the net electrical output of the plant and its overall thermal efficiency can be seen by studying Figure 6.7, which shows the layout of a typical power reactor.

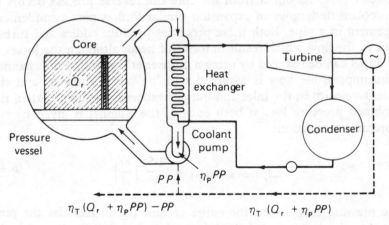

Figure 6.7. Layout of a typical power reactor

If the thermal output of the reactor is Q_r and the power to circulate the coolant is W or $\eta_\text{P}PP$, then the total thermal output of the plant is $(Q_\text{r} + \eta_\text{P}PP)$ and if losses are neglected this heat is transferred to the steam. If the combined thermal efficiency of the steam cycle and turbo-generator efficiency is η_T (this value is usually about 0·35 to 0·4), then the gross electrical output is:

$$\eta_\text{T}(Q_\text{r} + \eta_\text{P}PP)$$

and the net electrical output is:

$$\eta_T(Q_r + \eta_P PP) - PP$$

The overall thermal efficiency of the plant is:

$$\eta_{Net} = \frac{\text{Net electrical output}}{Q_r} = \eta_T - \frac{PP}{Q_r}(1 - \eta_T\eta_P)$$

The effect of pumping power on overall efficiency can be illustrated. If $PP = 0.05Q_r$ (a typical value for a gas-cooled reactor), and if η_T and η_P are 0·35 and 0·9 respectively, then the overall efficiency is 0·316. This is a 10 per cent reduction of the efficiency without any pumping requirement.

A useful criterion for gaseous coolants can be obtained by considering a reactor with a fixed thermal output, Q_r, fixed dimensions and fixed fuel and coolant temperatures. The coolant flow rate, specific heat and density depend on the identity of the coolant and its pressure. The pumping power can be written as:

$$PP = \frac{\dot{m}}{\eta_P \bar{\rho} \rho_1}\left(\frac{\dot{m}}{A}\right)^2\left\{\frac{4fL}{2d_e} + \frac{T_{c2} - T_{c1}}{\bar{T}_c}\right\} \tag{6.51}$$

The temperature term is constant and the friction factor f may be written, using Reynolds' Analogy, as:

$$f = 2(St) = \frac{2hA}{\dot{m}c_p} = \frac{2hA\,\Delta T_c}{Q_r} \tag{6.52}$$

In view of our assumption that the thermal output and temperatures in the fuel and coolant are fixed, h is constant and so also is the friction factor f.

Now we can write:

$$PP \propto \frac{\dot{m}^3}{\bar{\rho}\rho_1} \tag{6.53}$$

For a given thermal output and ΔT_c, $\dot{m} \propto 1/c_p$, and since coolant temperatures are fixed $\rho_1 \propto \bar{\rho}$, therefore:

$$PP \propto \frac{1}{c_p^3 \bar{\rho}^2} \tag{6.54}$$

This result indicates that the pumping power is inversely proportional to the square of the coolant density. For a gas the density can be increased by increasing the pressure, and this is done in gas-cooled reactors, the limit being imposed by the strength of the pressure vessel containing the core. For any given temperature and pressure, the densities of all gases are proportional to their molecular masses, consequently the preceding equation may be written as:

$$PP \propto \frac{1}{c_p^3 M^2} \tag{6.55}$$

where M is the molecular mass.

This result indicates that in a reactor in which the thermal output, dimensions, temperatures and gas pressure are fixed, the pumping power is least for the coolant whose value of $c_p^3 M^2$ is greatest, and this is a useful, though not infallible guide to the relative merits of gas coolants. Table 6.2 shows the values of $c_p^3 M^2$ for several gases referred to a value of 1 for hydrogen at 500 K.

Table 6.2 shows that from the point of view of heat transport and pumping power, hydrogen is an excellent coolant, however its strong chemical activity precludes its use. It may, however, find an application as the coolant-propellant in nuclear reactor powered space vehicles. After hydrogen, helium and carbon dioxide are the best gas coolants, with carbon dioxide showing an advantage at high temperatures.

The control of coolant flow in the outer channels of a reactor core by orifices or gags to increase the coolant outlet temperature has already been mentioned. This restriction in coolant flow is also desirable from

Table 6.2. Values of $c_p^3 M^2$ for gases

Gas	$c_p^3 M^2$	
	500 K	800 K
Hydrogen	1	1·04
Helium	0·181	0·181
Carbon dioxide	0·165	0·252
Air	0·075	0·091

the point of view of reducing pumping power. The core pressure drop is determined by the flow in the central channel, and the pressure drops in all other channels must be the same. The pumping power is therefore proportional to the mass flow of coolant, and if this mass flow is reduced by gagging the outer channels, the pumping power can be reduced. In the event that the mass flow in each channel is proportional to the thermal output of that channel, then the ratio of the thermal output in any channel to the power required to pump coolant through that channel is constant, and equal to the ratio of thermal output to pumping power for the whole core.

Chapter 7

Thermodynamic aspects of nuclear power plant

Up to the present date the majority of nuclear reactors built in the world have been designed for the generation of electric power. A few large reactors have been built for the production of plutonium, and many smaller reactors have been built for research, isotope production and as pilot plant for large power reactors. It is true, however, to say that the most important type of reactor is the power reactor and, as was pointed out in Chapter 3, the great effort devoted to the research and development of nuclear reactors can only be justified if it enables the world's resources of uranium and thorium to be used for power generation.

In a complete nuclear power plant the reactor is one component, the source of energy. Two other important components are the heat exchangers in which heat is transferred from the reactor coolant to the working fluid in the power cycle, and the power cycle itself in which the energy of the working fluid is converted to work in the turbines. In some types of reactor the coolant serves also as the working fluid in the power cycle and there are no heat exchangers. These are known as direct-cycle reactors.

In this chapter we will describe briefly some of the principal types of nuclear power reactors operating at the present time, and we will consider in more detail the thermodynamic aspects of nuclear power plant, i.e. the heat exchangers and the power cycle. The heat exchanger of a nuclear power plant is equivalent to the boiler of the conventional fossil-fuel fired plant, with the difference that the high temperature post-combustion gases of the conventional plant are replaced by the high temperature coolant in the nuclear power plant. The temperature of the reactor coolant is not as high as the post-combustion gases in a conventional boiler, so the power cycle for a nuclear reactor may differ in some respects from conventional power cycles.

It will be assumed that the reader is familiar with the laws and basic principles of thermodynamics, and with the analysis of simple power cycles such as the Rankine and Joule cycles for steam and gas turbines.

7.1   Nuclear power plant

It is appropriate at this point to give a brief description of a complete nuclear power plant. In preceding chapters the reactor has been considered from the point of view of reactor physics, heat transfer and heat transport. Now, before the thermodynamic aspects of reactors are discussed, a typical nuclear power plant will be described. There are of course many variations possible with different choices of fuel, moderator and coolant, and the following description refers to no particular type.

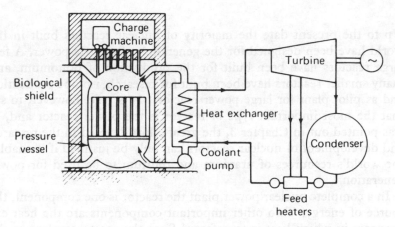

Figure 7.1.   The layout of a typical nuclear power plant

The core consists of an assembly of fuel and moderator, the fuel being cylindrical rods or plates in a lattice within the moderator. If the moderator is solid (graphite), the coolant flows in gaps or annular channels between the fuel and the moderator. If the moderator is liquid (H_2O or D_2O) it may also serve as the coolant and circulate through the core, or the coolant may be separate from the moderator and flow in pressure tubes between the fuel and the moderator.

The core is enclosed in a pressure vessel whose functions are to contain the coolant and support the core. In the case of pressurized and boiling water reactors, the system pressure may be as much as 160 bar to permit high coolant temperatures, while in gas-cooled reactors the gas pressure may be up to about 40 bar. The pressure vessel is usually a welded steel vessel; however, several recent reactors have been built with prestressed concrete pressure vessels lined with steel, in which case the pressure vessel also acts as the biological shield of the reactor.

The biological shield surrounds the reactor and prevents the escape of gamma radiation and neutrons which are dangerous to humans. (This topic is discussed in Chapter 9.) The biological shield of a power reactor is usually a concrete structure several feet thick, concrete being chosen for its structural characteristics, cheapness and the fact that it is sufficiently dense to be an effective shield for gamma radiation. Compact marine reactors use steel and water in preference to bulky concrete shielding.

Some reactors are designed so that the fuel can be loaded and unloaded from the core while the reactor is operating at power. In these reactors the charge–discharge machine is usually situated above the top shield of the reactor, and has access to the fuel elements in the core through standpipes in the top shield. Other reactors are designed so that the complete fuel in the core is unloaded at one time when the reactor is shut down, which involves removing the top cover of the pressure vessel. The control rod motors are usually placed above the reactor, and the control rods are driven into and out of the core in spaces between the fuel elements.

The coolant is circulated through the core and heat exchangers by pumps or blowers. The usual direction for coolant flow is upwards through the core. Normally there are four, six or even eight coolant loops and heat exchangers, so that one loop can be shut down for repair without affecting the operation of the reactor. The heat exchangers (sometimes called steam-raising units) may have separate economizer, evaporator and superheater sections, or they may be of the 'once-through' design.

The power cycle of a nuclear reactor is more or less conventional. All power reactors to date use steam power plant operating on the Rankine cycle; however, as pointed out above, steam cycles for nuclear plant may differ in some respects from conventional steam cycles because the temperatures generally available in reactors are less than in conventional boilers. Some of these variations will be described in this chapter.

7.2 Some thermodynamic preliminaries

The function of a thermodynamic power cycle is to convert heat to work. According to the Second Law of Thermodynamics, it is impossible to convert heat entirely and continuously into work, so the power engineer is concerned to convert as much as possible of the heat received in a thermodynamic cycle into work, while rejecting as little as possible to a heat sink. The thermodynamic efficiency of a power cycle is the ratio of the work output to the heat received.

The basic thermodynamic power cycle is the Carnot cycle. In this cycle all heat is received isothermally at a temperature T_1, and all heat rejected is rejected isothermally at a lower temperature T_2. All processes in the cycle are reversible, and its thermal efficiency is:

$$\eta_{\text{th}} = 1 - \frac{T_2}{T_1}$$

No practical power plant operates on the Carnot cycle, which for given values of T_1 and T_2 has the highest possible efficiency of any power cycle. However the expression for the Carnot cycle efficiency can be applied in a general way by observing that just as the efficiency of the Carnot cycle can be increased by increasing T_1 or decreasing T_2, so the efficiency of any power cycle can be increased by raising the average temperature at which heat is transferred to the working fluid, or reducing the average temperature at which heat is rejected by the working fluid in the cycle. In actual power cycles the minimum temperature at which heat is rejected is limited by the temperature of naturally occurring heat sinks, namely the earth's atmosphere and

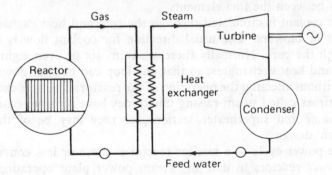

Figure 7.2. The layout of a gas-cooled reactor plant

oceans. This limits the minimum temperature of heat rejection to about 30°C. The improvement in the thermal efficiency thus becomes a matter of transferring heat to the working fluid at as high an average temperature as possible.

In a nuclear power plant the temperature at which heat is transferred to the working fluid depends on the fuel temperature, the temperature drop from the fuel to the coolant in the reactor, and the temperature drop in the heat exchangers between the reactor coolant and the working fluid of the power cycle. To study the influence of the heat exchangers on the overall efficiency of a nuclear power plant we will consider the gas-cooled reactor shown in Figure 7.2.

In this system the heat exchanger is an essential intermediate component between the reactor and the power cycle, which in this case is a simple Rankine cycle with H_2O as the working fluid. It is important that the temperature drop between the primary gas coolant and the

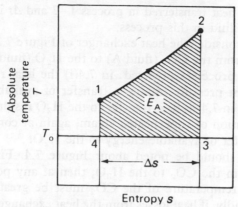

Figure 7.3.   Temperature-entropy diagram to illustrate available energy

H_2O in the heat exchanger should not be larger than necessary, as an increased temperature drop in the heat exchanger implies a lowering of the average temperature at which heat is transferred to the working fluid, with a consequent loss of efficiency in the power cycle.

A useful thermodynamic quantity in the discussion of heat exchangers and power cycles is available energy, E_A. The available energy is defined as the maximum energy transferred to a fluid in a heat transfer process which can be converted to work in a thermodynamic cycle. Consider the temperature–entropy diagram for some fluid, Figure 7.3. The fluid receives heat in a process during which its temperature varies along the path 1–2. (In a heat exchanger through which the fluid is flowing, such a process would be essentially a constant pressure process, and would follow a constant pressure line on the *T*–s diagram.)

The question arises as to what other processes, taken with the process 1–2 to form a thermodynamic cycle, will produce the maximum work. According to the Second Law of Thermodynamics, the maximum work can be obtained from a cyclic process provided that all heat rejected is rejected at the lowest possible temperature, namely the absolute temperature of the earth's atmosphere and oceans, T_0. Thus the 'maximum work' cycle in Figure 7.3 is completed by a reversible adiabatic (isentropic) expansion from 2 to 3 at temperature T_0, isothermal heat rejection at T_0 to 4 and finally reversible adiabatic compression from 4 to 1, thus completing the cycle.

The available energy of the heat transfer process 1–2

 = The work down in the cycle 1234,

 = (The heat received from 1 to 2) − (The heat rejected from 3 to 4),

 = $Q - T_0 \Delta s$,

where Q is the heat transferred in process 1–2 and Δs is the change of entropy of the fluid in this process.

Let us now consider the heat exchanger of Figure 7.2 in which heat is transferred from the CO_2 (fluid A) to the H_2O (fluid B). Figure 7.4 shows the two processes involved; in 7.4(i) the heat rejection by the CO_2 at constant pressure implies a transfer of available energy *from* the CO_2, while in 7.4(ii) heat transfer to the H_2O resulting in evaporation and formation of superheated steam, again at constant pressure, implies a transfer of available energy *to* the H_2O.

Two points should be noted about Figure 7.4. Firstly, if heat is transferred from the CO_2 to the H_2O, then at any point in the heat exchanger the temperature of the CO_2 must be greater than that of the H_2O. Secondly, if heat losses from the heat exchanger are negligible, then all the heat transferred from the CO_2 passes to the H_2O and the

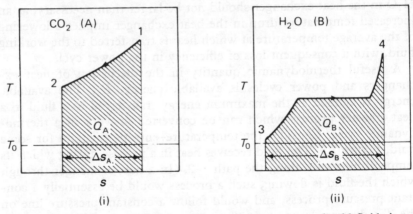

Figure 7.4. Temperature-entropy diagrams for the hot and cold fluids in a heat exchanger

area under the line 1–2 in Figure 7.4(i) is equal to the area under the line 3–4 in Figure 7.4(ii). It follows that the increase of entropy of the H_2O (Δs_B) is greater than the decrease of entropy of the CO_2 (Δs_A), and there is an overall increase of entropy of the two fluids passing through the heat exchanger.

The available energy transferred from the fluid A is equal in magnitude, but of opposite sign, to the available energy that would be transferred to it if it were being heated from 2 to 1 instead of being cooled from 1 to 2. Thus the available energy transfer from fluid A per unit time is:

$$E_A(A) = Q_A - T_0\,\Delta s_A$$

$$= Q_A - T_0\,\dot{m}_A\,\Delta s_A$$

where Q_A is the heat transfer from fluid A per unit time; $\dot{m}_A$ is the mass flow of fluid A per unit time; and Δs_A is the change of specific entropy of fluid A in the heat exchanger. (In the case of fluid A, both Q_A and Δs_A are of negative sign according to the usual thermodynamics sign convention.)

The rate of transfer of available energy to fluid B per unit time is (using the same symbols with suffix B),

$$E_A(B) = Q_B - T_0 \dot{m}_B \, \Delta s_B$$

(In this case both Q_B and Δs_B are positive.)

If heat losses from the heat exchanger are negligible, $Q_A = -Q_B$, and the rate of loss of available energy in the heat exchanger is:

$$E_A(\text{loss}) = E_A(A) + E_A(B)$$
$$= -T_0(\dot{m}_A \, \Delta s_A + \dot{m}_B \, \Delta s_B)$$

The term in brackets is positive because, as already noted, $|\Delta s_B| > |\Delta s_A|$. The negative sign for the quantity on the right of this equation indicates that there is a loss of available energy, and this represents the rate of loss of 'possible work' in the heat exchanger as a result of irreversible heat transfer.

The loss of 'possible work', which is valuable energy, can be reduced by reducing the temperature difference between the two fluids in the heat exchangers; however, any reduction in the temperature difference means that for a given reactor thermal output, the heat transfer surface, and hence the size and cost of the heat exchangers, must be increased. For any given set of economic factors, such as capital costs, fuel costs and interest rates, there is an optimum economic solution which determines the temperature difference between the primary and secondary fluids in the heat exchangers.

As an example, we will consider the gas-cooled reactor plant of the type shown in Figure 7.2, and apply some figures which are typical of the first generation of British reactors to determine the loss of available energy in the heat exchangers.

Carbon dioxide	*Inlet to heat exchangers*	*Outlet from heat exchangers*
Temperature (°C)	405	246
Enthalpy decrease, Δh (kJ/kg)	171	
Entropy decrease, Δs (kJ/kg K)	0.267	

Steam conditions at outlet from heat exchangers: 49 bar, 395°C,

$h = 3185$ kJ/kg, $s = 6 \cdot 639$ kJ/kg K
Feed water: 135°C, $h = 567$ kJ/kg, s $1 \cdot 687$ kJ/kg K
Datum temperature: $T_0 = 20$°C.

Figure 7.5 shows the temperature–heat transfer diagram for the heat exchangers based on 1 kg of H_2O. It is clear from the diagram that at all points in the heat exchangers the H_2O temperature is less than the corresponding CO_2 temperature, although at two points, the outlets of the economizer and the superheater, the H_2O temperature approaches quite closely the CO_2 temperature. It is also clear that because of the

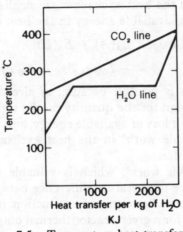

Figure 7.5. Temperature-heat transfer diagram

characteristic shape of the diagram, the average temperature difference between the H_2O and CO_2 is much greater than the minimum temperature difference, and the pressure at which steam is generated is very low compared with the steam pressure in modern conventional power plant which typically is about 160 bar.

The calculation is based on 1 kg of H_2O passing through the heat exchanger.
The heat transferred per kg of H_2O = $(3185 - 567) = 2618$ kJ/kg
The mass flow of CO_2 per kg of H_2O = $\dfrac{2618}{171} = 15 \cdot 3$ kg/kg of H_2O.

$\Delta s_{H_2O} = 6 \cdot 639 - 1 \cdot 687 = 4 \cdot 952$ kJ/kg K

$\Delta s_{CO_2} = 0 \cdot 267$ kJ/kg K

Loss of available energy in the heat exchanger/kg of H_2O:

$= -293 \, [4 \cdot 952 - (15 \cdot 3 \times 0 \cdot 267)]$
$= -254 \cdot 0$ kJ/kg

Available energy transferred to H_2O in heat exchanger:

= 2618 − (293 × 4·952)

= 1167 kJ/kg

In the turbines the work done/kg of steam = 792 kJ, and the loss of available energy in the thermodynamic power cycle due to irreversible expansion in the turbine and irreversible heat rejection in the condenser is 1167 − 792 = 375 kJ/kg of steam. Evidently the loss of available energy in the heat exchangers is almost one-third of the actual work done, which explains the need for careful design of the heat exchangers in this type of reactor.

7.3 Heat exchangers and steam cycles for gas-cooled reactors

The layout of a possible heat exchanger for a gas-cooled reactor is shown in Figure 7.6(i) in which steam is generated at one pressure in a single-pressure cycle. If the CO_2 inlet and outlet temperatures at the heat exchanger T_{c1} and T_{c2}, the minimum CO_2–H_2O temperature differences in the economizer and superheater ΔT_{ec} and ΔT_{sup}, and the feed-water temperature T_{fd} are all known, it is possible to determine the maximum pressure and temperature at which steam may be generated by the following method.

The final temperature of the superheated steam, T_s, is equal to the gas inlet temperature minus the minimum temperature difference at the superheater outlet, i.e. $T_s = T_{c1} - \Delta T_{sup}$. Although the pressure and saturation temperature at which steam is generated are not known, reasonable first approximations for both quantities can be made by assuming that $T_{sat} = T_{c2}$, and based on this estimate, values h_f the specific enthalpy of saturated water leaving the economizer, and h_s the specific enthalpy of superheated steam leaving the heat exchanger can be determined. The specific enthalpy of the feed water is determined by its temperature.

The ratio of the heat transferred in the economizer to the heat transferred in the whole heat exchanger can be expressed either in terms of the H_2O enthalpies (using trial values as explained above) or in terms of CO_2 temperatures as:

$$\frac{\text{Heat transfer in the economizer}}{\text{Heat transfer in the whole heat exchanger}} = \frac{h_f - h_{fd}}{h_s - h_{fd}}$$

$$= \frac{C_{pg}(T'_c - T_{c2})}{C_{pg}(T_{c1} - T_{c2})}$$

where C_{pg} is the specific heat of CO_2 and T'_c is the temperature of the CO_2

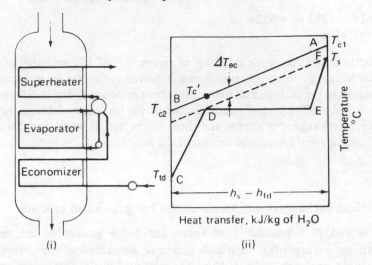

Figure 7.6. The heat exchanger and pinch point diagram for a gas-cooled reactor with a single-pressure steam cycle

at the point where it passes from the evaporator into the economizer. If C_{pg} is assumed to be constant:

$$\frac{h_f - h_{fd}}{h_s - h_{fd}} = \frac{T'_c - T_{c2}}{T_{c1} - T_{c2}}$$

from which:

$$T'_c = T_{c2} + \frac{(T_{c1} - T_{c2})(h_f - h_{fd})}{h_s - h_{fd}}$$

A second value of T_{sat} can now be found as:

$$T_{sat} = T'_c - \Delta T_{ec}$$

This second value of T_{sat} should be compared with the trial value used at the start of the calculation. If the two are in agreement, then the original approximation was correct; if not, the calculation can be repeated using the second value of T_{sat} as the starting point for the second iteration.

The temperature-heat transfer diagram can be constructed from the results of this calculation, and is shown in Figure 7.6(ii). This diagram is called a pinch-point diagram, the pinch-points being the two points at which the temperatures of the two fluids approach each other most closely, at outlet from the economizer and superheater. It is clear from the pinch-point diagram and the previous example that even with careful design the pressure and temperature at which steam is generated may be very low and the loss of available energy may be significant if steam is generated at a single pressure in the simple heat exchanger shown in Figure 7.6(i).

For example, taking some figures from the Bradwell power station (which is similar to Calder Hall and was completed in 1962) the CO_2 temperatures are $T_{c1} = 390°C$ and $T_{c2} = 175°C$, the pinch point temperature differences in both economizer and superheater are about 17°C, and the feed water temperature $T_{fd} = 86°C$. It can be shown using the method described above that if steam is generated in a single-pressure cycle, the conditions at the superheater outlet would be 15·5 bar pressure and 373°C. These are far below the standard conditions in modern steam power plant.

Two further points are evident from the pinch point diagram, Figure 7.6(ii). Firstly, the pressure at which steam is generated can be raised if the gas temperature at outlet from the heat exchanger is raised; however, this is in direct conflict with the condition necessary to reduce the pumping power. Secondly, the pressure at which steam can be generated is lowered if feed-heating is employed. The effect of feed-heating is to raise point C in Figure 7.6(ii), with the result that, since the gradient of the line CD is almost unchanged, point D moves slightly downwards and the evaporation temperature and pressure are slightly reduced.

In order to reduce the loss of available energy and generate some of the steam at a higher pressure than is possible in a single-pressure cycle, a dual-pressure cycle may be used, and all except one of the eleven nuclear power stations completed in Britain between 1956 and 1968 have used dual-pressure cycles. The layout of a typical dual-pressure cycle power plant and its pinch point diagram are shown in Figure 7.7(i) and (ii) which refers (with minor modifications) to the Bradwell power station. The principal difference between single- and dual-pressure cycles is in the heat exchangers, where separate economizers, evaporators and superheaters have to be provided for high-pressure (H.P.) and low-pressure (L.P.) steam. The arrangement of these components in the heat exchanger may vary from one design to another. In the power cycle, high-pressure steam is supplied to the high-pressure stages of the turbine and then mixes with steam from the low-pressure superheater before passing to the low-pressure stages of the turbine.

In the Bradwell station one third of the steam is generated at 15·5 bar pressure and two thirds at 54 bar pressure with a common final temperature of 373°C. The gross electrical output from two 531 MW(th) reactors is 350 MW, which is about 20 MW higher than it would be if all steam were generated at 15·5 bar pressure.

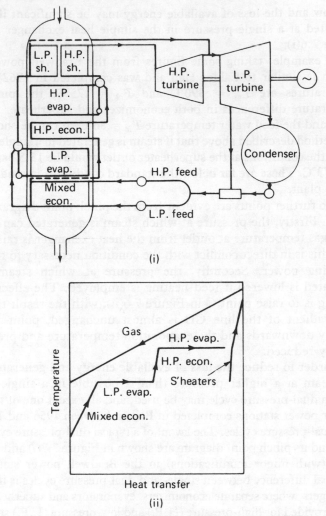

Figure 7.7.   The layout of a dual-pressure steam cycle, gas-cooled nuclear power plant

7.4 The first generation of British gas-cooled reactors

The early development of nuclear reactors in Britain after the Second World War was determined by a number of factors, one was the need to produce plutonium for nuclear weapons, another the fact that only natural uranium (and not enriched uranium) was available to fuel these reactors, and thirdly heavy water was not available to be used as the moderator of these reactors.

Table 7.1. *Design characteristics of four Mark 1 gas-cooled nuclear power stations*

	Calder Hall	Bradwell	Oldbury	Wylfa
Date of commissioning	1956	1962	1968	1969
Net electrical output (MW)	156	350	560	1180
Gas temp. at inlet to reactor (°C)	140	180	245	246
Gas temp. at outlet from reactor (°C)	334	390	410	405
Gas pressure (bar)	7·9	10	24·8	27·6
Steam conditions bar, (°C) % of flow				
Low-pressure	4·3, 177, 23	15·5, 373, 32	48·5, 394, 40	48·3, 396
High-pressure	14·5, 313, 77	54, 373, 68	95, 394, 60	
Feedwater temp. (°C)	38	86	138	135
Reheat	None	None	Full reheat in heat exchangers	Steam—steam reheater
Net thermal effy. (%)	21·5	28	33·6	31·4

Consequently early development in Britain was based on natural uranium fuelled, graphite moderated reactors. Gas cooling was chosen and, with the exception of the first reactors built at Windscale specifically for the production of plutonium, carbon dioxide was selected as the most suitable coolant. The fuel cladding material chosen was an alloy of magnesium called Magnox.

This choice of materials characterized the first generation of British nuclear power reactors whose prototypes were the Calder Hall reactors, commissioned in 1956 and designed for the production of plutonium and the generation of electricity. The reactors based on this design, of which 26 were built in 11 power stations, have been known as the Calder Hall, Magnox or Mark 1 reactors.

There was a progressive development of these reactors in the years between 1956 and 1970, and Table 7.1 shows design figures for four of these reactors to illustrate this. One of the most notable improvements was the increasing CO_2 pressure made possible by increasing pressure vessel thickness, and ultimately the introduction of prestressed concrete pressure vessels acting as combined pressure vessel and biological shield.

One of the most important limitations in the design of these reactors was the maximum operating temperature of the Magnox cladding, which was considered to be about 450°C. This in turn led to a maximum CO_2 temperature at the reactor outlet of 400 to 410°C. Later, during the

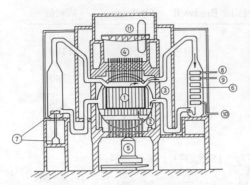

1. Core
2. Pressure vessel and support structure
3. Biological shield
4. Control rods and drive motors
5. Fuel charge and discharge machine
6. Heat exchanger (8 per reactor)
7. Coolant gas circulators and drive motors
8. High pressure steam supply
9. Low pressure steam supply
10. Feed water supply
11. Travelling crane and service machine
12. Valves in hot and cold gas ducts

Figure 7.8.   Layout of the core and heat exchangers of Hunterston A, a British Mark 1 gas-cooled reactor

operating life of these reactors, it was found that at this temperature the CO_2 caused corrosion of certain mild steel components in the coolant pipework, and in order to restrict this the maximum CO_2 temperature was lowered to about 370°C. Consequently these reactors have been operating at lower temperatures and powers than their design figures.

The layout of a typical Magnox reactor is shown in Figure 7.8 which shows the Hunterston A power station. The graphite core, 7·3 m high by 13·6 m diameter is enclosed in a spherical pressure vessel 21·3 m diameter. The core has 3248 channels, each containing 10 fuel elements stacked one above the other. The fuel is in the form of natural uranium rods 2·9 cm diameter by 62 cm long enclosed in finned Magnox tubes. The cross-section of these is shown in Figure 7.9. The total mass of uranium per reactor is 251 tonnes.

The CO_2 coolant is circulated at a pressure of 11 bar with inlet and outlet temperatures of 160°C and 370°C respectively (operating values, not design values). The thermal output of each reactor is 630 MW and the electrical output of the power station, which has two reactors, is 300 MW. The Magnox reactors are designed to be refuelled on load, and Hunterston A differs from other reactors of the same type in that its refuelling machine is located under the core and not above it, as is the case with all other Magnox reactors.

Now, in 1988, the earliest of the Magnox reactors are nearing the end of their design lives and are the subject of inspection to ascertain if they can continue to operate, possibly at reduced power, for a further few years. It seems certain that once the existing Magnox reactors come to the end of their lives no others of the same design will be built, and their characteristic design features such as metallic uranium fuel and dual-pressure steam cycles will not be seen again in future reactors. However, the operating record of the nine Magnox power stations operated by the electricity generating boards in the UK has been good. For example, Hunterston A

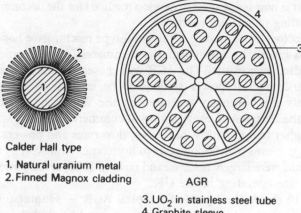

Calder Hall type
1. Natural uranium metal
2. Finned Magnox cladding

AGR
3. UO$_2$ in stainless steel tube
4. Graphite sleeve

Figure 7.9. Fuel elements of British gas-cooled reactors

has had one of the best lifetime load factors among the world's nuclear power stations; a tribute to its reliability.

7.5 Advanced gas-cooled reactors

The steam conditions attainable in all the British Magnox nuclear power stations up to and including Wylfa are well below the steam conditions in modern conventional fossil fuel power plant, in which 160 bar pressure and 565°C are typical figures. Further advances in gas-cooled reactors to enable these figures to be achieved can only be made by changing from Magnox cladding to materials suitable for high temperature use such as stainless steel or Zircaloy, and by changing from metallic uranium fuel to uranium oxide or uranium carbide, compounds which can be fabricated as ceramics with extremely high melting points and no phase change problems. In addition, the use of better types of steel is necessary to eliminate the CO$_2$ induced corrosion mentioned in the preceding section. Increased fuel and cladding temperatures enable the rating of the fuel to be increased, which in turn makes possible a greatly increased thermal output as compared with a Magnox reactor of the same size.

The development of the advanced gas-cooled reactor (AGR) started in Britain in the late 1950s, and the prototype AGR was built at Windscale, Cumbria by the United Kingdom Atomic Energy Authority and commissioned in 1962. This reactor used uranium oxide (UO$_2$) fuel in stainless steel cladding, the fuel containing enriched uranium with 3·1 per cent ^{235}U. The moderator and coolant were graphite and carbon dioxide respectively. The thermal power of the reactor was 105 MW, and the electrical power was 33 MW, giving a thermal efficiency of 31 per cent. The Windscale AGR operated successfully for 20 years and has recently been shut down. It is now being used to develop methods for the decommissioning and dismantling of nuclear reactors.

The satisfactory performance of this prototype reactor gave hope that the design could readily be extrapolated to commercial power station size of several hundred megawatts. The AGR was chosen for the second generation of British nuclear power stations to be introduced in the 1970s. Unfortunately the optimism generated by the Windscale AGR was not justified as the design and construction of commercial AGRs, with larger sizes and higher temperatures, pressures, flow rates and powers, encountered many problems which took years to overcome. However, by the 1980s these problems were largely resolved and now, in 1988, seven power stations of this type are operating in the UK.

Figure 7.10 shows the layout of a typical AGR – Hunterston B. The graphite core is $8 \cdot 2$ m high by $9 \cdot 3$ m diameter and has 308 channels, each one containing 8 fuel elements stacked one on top of the other. Each fuel element, $1 \cdot 036$ m long by $0 \cdot 264$ m diameter, contains 36 fuel pins $14 \cdot 5$ mm diameter enclosed in a graphite sleeve as shown in Figure 7.9. The pins consist of stainless steel tubes enclosing uranium oxide pellets with uranium enriched to $2 \cdot 0$ to $2 \cdot 6$ per cent ^{235}U. The total mass of uranium in the core is $122 \cdot 5$ tonnes.

The carbon dioxide coolant circulates at the rate of 3662 kg/s per reactor,

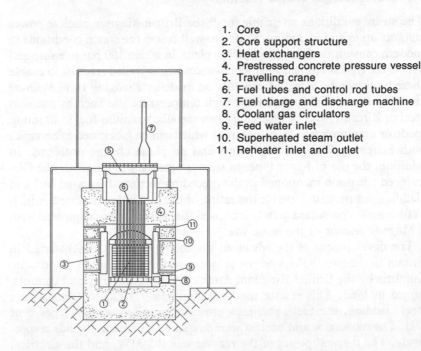

1. Core
2. Core support structure
3. Heat exchangers
4. Prestressed concrete pressure vessel
5. Travelling crane
6. Fuel tubes and control rod tubes
7. Fuel charge and discharge machine
8. Coolant gas circulators
9. Feed water inlet
10. Superheated steam outlet
11. Reheater inlet and outlet

Figure 7.10.   Layout of the core, heat exchangers and pressure containment of the Hunterston B advanced gas-cooled reactor

with inlet and outlet temperatures of 316°C and 654°C, and steam is generated in the heat exchangers at 167 bar pressure, 541°C with reheat at 42 bar pressure to 541°C. The heat exchangers are located in the annular space between the core and the inside of the prestressed concrete pressure vessel.

The thermal output per reactor is 1496 MW and the net electrical output of the power station, which has two reactors, is 1248 MW, giving a net thermal efficiency of 41·7 per cent; considerably higher than the efficiency of the Magnox power stations.

The AGR programme in Britain has suffered from long time-delays in construction and rather disappointing operational performance, typified by the low load factors of some of the power stations. This experience, together with the high capital cost of this type of reactor as compared with the pressurized water reactor must count against the continuation of the AGR programme, and it is quite possible that no more power stations of this type will be built.

7.6 High temperature gas-cooled reactors

Further advances in gas-cooled reactor technology have led to the high temperature gas-cooled reactor (HTGR), of which only five have been built to date. It is proposed that this type of reactor will operate as a thermal breeder reactor using the thorium 232/uranium 233 breeding process, see Section 3.4. Early HTGRs have been fuelled with a mixture of ^{232}Th and highly enriched uranium containing about 93 per cent ^{235}U. In future the fissile material will be ^{233}U once sufficient quantities of this isotope have been created by breeding.

The thorium—uranium fuel is in the form of the carbides ThC_2 and UC_2, and very small spherical particles of this mixture less than 1 mm diameter are coated with pyrocarbon to retain the fission products. The coated particles, about 1 mm diameter, may be mixed with graphite to form a homogeneous fuel-moderator mixture. Because of the very high melting points of ThC_2, UC_2 and graphite, these fuel-moderator mixtures can operate at very high temperatures, and coolant temperatures in the range 750 to 1000°C are possible. The most suitable coolant for the HTGR is helium because it is chemically inert, has good heat transfer properties and low neutron absorption. Conventional cladding materials such as stainless steel are not required, and the use of low neutron capture materials such as graphite and helium results in good neutron economy and enhances the breeding ratio.

At the present time the HTGR has not been widely adopted. A European cooperative venture, the DRAGON project at Winfrith in England, showed the feasibility of this design 20 years ago, but the project was abandoned without going beyond the stage of an experimental small-scale HTGR.

Some interest has been shown in America in the HTGR, and the Fort St Vrain power station in Colorado is of this type. The coated particle fuel, containing a mixture of ThC_2 and UC_2 is loaded into holes in graphite blocks which form the core of the reactor. The fuel loading of the reactor is $19 \cdot 48$ tonnes of thorium and $0 \cdot 88$ tonnes of highly enriched uranium. The maximum fuel temperature is 2300°C and the helium temperature at outlet from the reactor is 770°C. The thermal power of this reactor is 842 MW, the electrical power is 330 MW and the thermal efficiency $39 \cdot 2$ per cent.

The pebble-bed reactor (AVR) is a unique type of high temperature reactor which has been built at Jülich in West Germany. The UC_2/ThC_2 particles are dispersed in graphite, and this homogeneous fuel-moderator matrix is fabricated in the form of spheres 28 mm diameter. These spheres are loaded into the reactor vessel to create a critical mass and helium is blown upwards through the spaces between the spheres. The reactor may be refuelled on load by removing irradiated fuel spheres from the bottom of the reactor vessel and replacing them by new ones. Gas outlet temperatures of 900°C have been achieved in this reactor, which makes it suitable for high temperature industrial uses.

The very high coolant temperatures which are possible in the HTGR make it possible to generate steam at conditions equivalent to those in modern conventional power plants. However, this high temperature can be used more effectively in a gas turbine than in a steam turbine, as the latter does not benefit greatly from coolant temperatures higher than those available in AGRs. Gas turbines, on the other hand, require very high turbine inlet temperatures to be competitive, and a combination of the HTGR with a direct-cycle gas turbine may be a promising system, although nothing like this has yet been built. If such a power system is built, helium will prove not only to be an ideal coolant for the HTGR, but also an ideal working fluid in the gas turbine because it has good thermodynamic properties. For example, in an ideal Joule cycle in which the maximum and minimum temperatures are fixed, helium can produce a higher specific power with a lower pressure ratio than either CO_2 or air.

7.7 Pressurized water reactors

The excellent properties of water as the moderator for a thermal reactor, e.g. the good slowing down characteristics and the short diffusion and slowing down lengths, makes it possible for a water moderated reactor to be very much smaller than a graphite moderated reactor. A water moderated reactor may have core dimensions of only 2 or 3 metres height and diameter. This fact was realized by, among others, the American naval captain Rickover in the late 1940s after the Second World War. He appreciated the great strategic advantage that nuclear power plant would give to naval vessels, particularly submarines, which would be able to cruise underwater for prolonged periods. The earliest development of water moderated reactors was directed towards producing power plants for submarines, and the first nuclear powered submarine, USS Nautilus, went to sea in 1955.

Another important characteristic of water as the coolant for a nuclear reactor is its saturation temperature and pressure relationship — the fact that if water is to be prevented from boiling at high temperatures (in the range 100 to 370°C) it must be maintained at a correspondingly high pressure. From the thermodynamic point of view it is desirable that if water is used as the coolant in a power reactor, its temperature should be as high as possible, and its pressure must therefore also be high — hence the name pressurized water reactor (PWR) for this type of reactor in which water acts as both the coolant and moderator.

In view of the rather high capture cross-section of water (0·66 barns/molecule) it is necessary to use slightly enriched uranium as the fuel. The enrichment necessary is typically 2·5 to 3 per cent. Finally, in order to prevent corrosion caused by the high temperature water, stainless steel or zirconium alloy cladding is required, these alloys having the necessary corrosion resistance.

The PWR is therefore characterized by being much more compact than a graphite moderated reactor and having a higher power density, as measured by the power output per unit volume of core. The water acting as moderator and coolant is at a high temperature and very high pressure, typically 300°C and 160 bar. The pressure vessel containing the core and the primary loop pipework must be strong enough to withstand this very high pressure without the possibility of failure. Typical pressure vessel thickness is 215 mm. The compact size of the reactor leads to lower capital costs than for graphite moderated reactors.

The development of PWRs for electricity generation started just after the early development for submarine propulsion, and in 1957 the Shippingport, Pennsylvania power station was commissioned. Its electrical output was 75 MW, and the designer and manufacturer was the Westinghouse Electric Corporation, a company that has remained at the forefront of PWR development. In the following 20 years the PWR was developed in America and large numbers of nuclear power stations of this type were built. PWRs were also exported to many countries and built under licence in France, West Germany and Japan. This development was interrupted but not halted by the accident to a PWR at Three Mile Island, Pennsylvania in 1979 (see Section 11.4), and in the last nine years increasing attention has been paid to the safety of this type of reactor.

After 30 years of development the PWR has now reached a stage at which most new reactors are similar to each other as regards their principal features

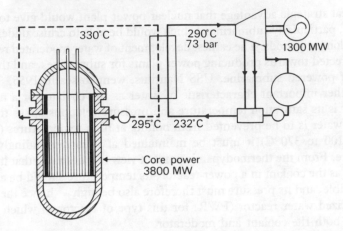

Figure 7.11.   Layout of the core and pressure vessel, and simplified steam
cycle of a pressurized water reactor

and operating parameters. The following description and the layout illustrated in Figure 7.11 can be regarded as typical of present PWRs rather
than as referring to any particular one.

The fuel is UO_2, enriched to about 2·5 per cent of ^{235}U, in the form of
pellets 8 mm diameter by 10 mm long, enclosed in 4 m long tubes of
Zircaloy-4 alloy 10 mm outside diameter. These tubes are closely packed
in square arrays of 20 cm side, there being about 236 tubes per fuel
assembly. These square fuel assemblies are arranged side by side to form
the core of the reactor, whose dimensions are approximately 3·8 m
diameter and 3·7 m high, within the pressure vessel which is 4·7 m inside
diameter and 10 m high. The water pressure in the core and pressure
vessel is 153 bar and the water temperatures at core inlet and outlet are
295°C and 330°C respectively. In the heat exchangers steam is generated
at 73 bar with a moisture fraction of less than 0·0025.

The present standardized PWR has a thermal power of 3800 MW, and
the electrical output is 1300 MW, giving a thermal efficiency of 34 per cent.
By comparison with gas-cooled, graphite moderated reactors, the PWR has
lower thermal efficiency and lower breeding ratio, but at a time when there
are abundant uranium supplies at not too high prices these factors are less
important than the lower capital cost of the PWR, which is now the most
widely used reactor in the world. See Appendix 4 for a summary of world
installed nuclear power.

Within the last decade in the UK the Central Electricity Generating Board
has, after 20 years of operating gas-cooled reactors, sought Government
approval to build a PWR. After a prolonged public enquiry the UK

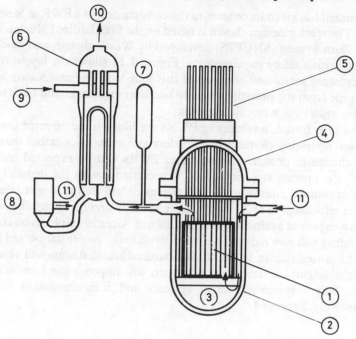

Figure 7.12.   Layout of the core, pressure vessel and one primary coolant loop
of the Sizewell B pressurized water reactor

1. Core
2. Pressure vessel
3. Core support structure
4. Upper hemispherical dome of the pressure vessel
5. Control rod drives
6. Heat exchanger
7. Pressurizer
8. Coolant water pump
9. Feed water inlet to heat exchanger
10. Saturated steam outlet from heat exchanger
11. Primary coolant return from pump to reactor

Sizewell B data

Thermal power: 3411 MW   Net electrical power: 1110 MW
Net thermal efficiency: 32·5 per cent
Fuel: UO$_2$ pellets   Enrichment (feed): 3·2 per cent   Fuel rod length: 3·66 m
Core height: 3·66 m   Equivalent core diameter: 3·37 m
Number of fuel assemblies: 193
Number of fuel rods per assembly: 264
Pressure vessel: height 13·55 m; inside diameter 4·39 m; thickness 215 mm
Coolant pressure: 156 bar   Coolant temperatures: inlet 293°C, outlet 325°C
Coolant flowrate: 18 470 kg/s
Steam supply conditions: 69 bar, 285°C, 0·9975 dryness fraction
Feedwater temperature: 227°C   Steam flowrate: 1908 kg/s

Government has given its consent to the construction of a PWR at Sizewell, Essex. The reactor design chosen is based on the Standardized Nuclear Unit Power Plant System (SNUPPS) developed by Westinghouse, and modified to meet British safety requirements. Figure 7.12 shows the layout of the core, pressure vessel and one loop of this four-loop design. (A loop refers to the pipe from the pressure vessel to one heat exchanger and back to the pressure vessel via a circulating pump.)

In the near future, advances in PWRs are likely to be directed towards improved operating characteristics, safety and economics rather than any major change in design. For example, PWRs will be expected to 'load follow', i.e. operate at variable load in accordance with the demand, as a matter of routine. Improved fuel management will lead to longer intervals between refuelling (18 or 24 months instead of the present 12 months) which will give improved availability and load factors. Greater emphasis on cooling water purity will give reduced radiation levels in the power station and lower doses to power station workers. Standardization of design will result in shorter construction times which in turn will improve the economics of PWRs. Safety systems, control systems and instrumentation will be progressively improved.

7.8 Boiling water reactors

There are obvious advantages in allowing the water in a water-cooled and moderated reactor to boil in the core of the reactor, particularly if the steam generated in the core, after being separated from the saturated water, is passed directly to the turbines in a direct-cycle as shown in Figure 7.13. This system eliminates the need for the heat exchangers of the PWR with their inevitable thermodynamic losses and increased cost. Furthermore, since no attempt is made to suppress boiling, the pressure in the reactor is much less than the pressure in a PWR which produces steam at the same conditions.

Early uncertainties associated with the boiling water reactor (BWR) concerned the effect on the safety and stability of the reactor of boiling in the core, and the hazard of radioactive contamination in a direct-cycle when steam generated in the reactor is circulated through the turbines. Experience has shown that these two factors do not present serious problems. Boiling in the core has been shown to be safe, and steam dryness fractions of up to 15 per cent at the core outlet have been achieved. The radioactive contamination problem can be largely overcome by ensuring a very high degree of purity of the water in the system.

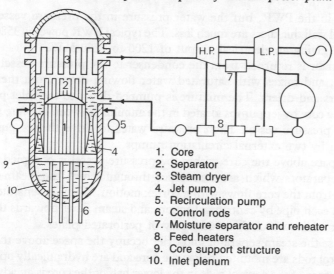

1. Core
2. Separator
3. Steam dryer
4. Jet pump
5. Recirculation pump
6. Control rods
7. Moisture separator and reheater
8. Feed heaters
9. Core support structure
10. Inlet plenum

Figure 7.13.   Layout of a direct-cycle boiling water reactor

The BWR has been developed in parallel with the PWR in the United States, and like the PWR it has been built in large numbers, both there and in other countries such as Sweden, West Germany and Japan. The earliest reactor of this type, the Vallecitos BWR with a power of 5 MW electrical, was commissioned in 1957. Since then nearly 40 power stations of this type have been built in the United States by the General Electric Company, the leading designer and constructor of this type of reactor.

In many respects the BWR is similar to the PWR. The main difference is the absence of a heat exchanger between the reactor and the power cycle, and there are other differences within the pressure vessel. The following data is typical of present BWRs:

The fuel pellets of UO_2 are 10·6 mm diameter by 12 mm long and are loaded into Zircaloy-2 fuel cladding tubes. These fuel rods are assembled in 7×7 or 8×8 arrays in boxes 14 cm square which make up the core. There are about 580 of these fuel boxes and a total of 140 t of fuel in the core which is 4·7 m diameter by 3·75 m high. The spacing between the fuel rods is slightly larger than in the PWR, giving a slightly larger core diameter. The uranium in the fuel is enriched to 1·7 to 2·5 per cent ^{235}U; replacement fuel has 2·5 to 3·1 per cent ^{235}U. The pressure vessel is 21·6 m high by 6·05 m inside diameter, and the wall thickness is 152 mm.

The boiling water coolant in the pressure vessel is at a pressure of 72·5 bar, and saturated steam is supplied to the turbines at 65 bar, 281°C. It may be noted that the steam supply conditions to the turbine are similar to those in the PWR, but the water pressure in the pressure vessel, and the vessel wall thickness are much less. The typical BWR power is 3580 MW thermal, giving an electrical output of 1200 to 1260 MW.

Feedwater is returned from the condenser to the pressure vessel above the core, and mixes with saturated water flowing down from the steam separators and driers. The mixture is pumped down to the inlet plenum below the core by jet pumps located in the annulus between the core shroud and the pressure vessel. High pressure water to drive the jet pumps is provided by two external circulation pumps.

The space above the core within the pressure vessel is occupied by the steam separators which are standpipes through which the steam—water mixture from the core flows with a vortex motion. Water is separated outwards in each pipe by centrifugal action, and steam flows upwards through the steam drier which is an assembly of perforated plates.

Because the steam separator and drier occupy the space above the core, the control rods are mounted below the core and are hydraulically operated. The presence of the control rods in the lower half of the core is an advantage as in that position they produce a more uniform axial power distribution in the core than would be the case if they were inserted at the top of the core.

BWRs are designed so that peak heat fluxes at rated operating conditions are approximately 50 per cent of the critical heat fluxes (see section 6.6). They also have negative void coefficients of reactivity (see section 8.6), which implies that as the reactor power increases and more boiling occurs with increased steam voidage, the reactivity of the reactor decreases and its power tends to drop. This is a safe and self-stabilizing effect, and it means that BWRs can be controlled during load-following by varying the coolant flow rate and hence the voidage in the core due to steam production. For example, if a power decrease is required, the coolant flow rate is decreased and the amount of steam voidage increases; this causes a power decrease without the need for control rod movement.

Advances in BWR design to produce the advanced boiling water reactor (ABWR) are mainly aimed to give operating and safety improvements. The objectives are to improve the load-following ability of the reactor, improve its safety and increase its availability and load factor while at the same time reducing operator radiation exposure levels and operating costs. Improvements include increased power output, internal recirculation pumps, improved fuel design, reinforced concrete containment and improvements in safety systems, control systems, instrumentation and radioactive waste handling. Core size is increased to 5·16 m diameter, pressure vessel internal diameter and thickness are 7·1 m and 174 mm respectively and the number of fuel assemblies is increased to 872. The reactor power is increased to 3926 MW thermal, 1356 MW electrical. In the steam cycle there are two stages of moisture separation and reheat between the high, intermediate and low pressure turbines.

7.9 Heavy water moderated reactors

The very low neutron capture cross-section of heavy water, and the low mass number of deuterium make heavy water an excellent moderator for thermal reactors, and it can also be used as the coolant. Heavy water moderated and cooled reactors can be fuelled with natural uranium and have good neutron economy. These factors lead to low fuel costs, good breeding ratios and high burnups. One disadvantage of heavy water is its very high cost which makes it important to avoid losses from pipes and heat exchangers. Another disadvantage is the critical temperature limitation which (just as in PWRs) limits the maximum coolant temperature and makes necessary a very high coolant pressure. The thermodynamic characteristics of heavy water cooled reactors and pressurized water reactors are very similar.

At the end of the Second World War Canada alone had the industrial plant necessary to produce substantial amounts of heavy water, and made use of this capacity to develop the heavy water reactor. In the last 40 years eighteen reactors of this type, known as CANDU reactors (Canadian deuterium uranium) have been built and they form the whole of Canada's nuclear power capacity. A small number of this type of reactor has also been exported, for example to India, Romania and Argentina.

The CANDU reactors have a number of features which distinguish them from other types. The moderator is contained in a cylindrical $26 \cdot 8$ mm thick stainless steel vessel whose axis is horizontal, called the calandria. Its dimensions are 6 m long by $7 \cdot 1$ m diameter. (All data given here refer to the Pickering, Ontario nuclear power station). 380 horizontal Zircaloy tubes pass through the calandria. Inside each of these calandria tubes there is a 104 mm inside diameter Zircaloy pressure tube, with a helium filled gas gap between the two tubes to provide thermal insulation. This allows the heavy water in the calandria to be maintained at a lower temperature (65°C) and pressure than the heavy water coolant within the pressure tubes. Thus the moderator and the coolant, though both heavy water, are physically separated in the reactor.

Inside each pressure tube there are 12 fuel bundles, each one $0 \cdot 5$ m long and consisting of an assembly of 28 fuel rods. The fuel is natural uranium oxide, UO_2, the fuel pellets are $22 \cdot 3$ mm long by $14 \cdot 8$ mm diameter, and are enclosed in Zircaloy-4 tubes $0 \cdot 42$ mm thick to make the $15 \cdot 6$ mm diameter fuel rods. The total fuel loading in the core is $90 \cdot 5$ t of UO_2, and the effective core diameter is $6 \cdot 74$ m. The maximum cladding surface temperature is 304°C, the maximum fuel temperature is about 2000°C and the maximum fuel rod linear heat rating is $52 \cdot 8$ kW/m.

The heavy water coolant pressure in the reactor is $88 \cdot 3$ bar, and the inlet and outlet temperatures are 250°C and 293°C respectively. In the heat exchangers steam is generated at 41 bar pressure, 251°C. The thermal power of each reactor (there are 8 at Pickering) is 1744 MW, and the net electrical output is 515 MW, giving a thermal efficiency of 29·5 per cent. However, this low efficiency is offset by the low fuel costs mentioned above.

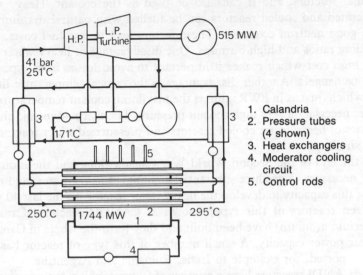

Figure 7.14.   Layout of the calandria, heat exchangers and simplified steam cycle of the CANDU reactor (Pickering)

An alternative version of the CANDU reactor is the heavy water-moderated, light water-cooled reactor. The replacement of the heavy water by light water as the coolant not only reduces the capital cost of the reactor, but also eliminates the financial penalty of leakage of expensive heavy water from the reactor coolant circuit. It is possible to use natural uranium fuel provided the quantity of light water is limited, however some reactors of this type use slightly enriched uranium. A further advantage of this system is that the light water can be allowed to boil in the pressure tubes, and a direct-cycle system can be used, with steam separation in external drums and dry saturated steam supplied to the turbines. This type of reactor combines the good features of heavy water moderation and the direct-cycle of the BWR.

The British steam-generating heavy water-moderated reactor (SGHWR) is an example of this direct-cycle system. The calandria is similar to that of the CANDU reactors; however, its axis is vertical. The pressure tubes, insulated from the calandria tubes by annular gas gaps, contain assemblies of 36 fuel rods consisting of 1·5 cm diameter pellets of UO_2 (2·28 per cent ^{235}U) in Zircaloy tubes. The pressure of the light water coolant in the pressure tubes is 67 bar, and the dryness fraction at outlet from the core is 11 per cent. After separation, dry saturated steam at 61·5 bar pressure is supplied to the turbines in a conventional cycle with reheating, moisture separation in the turbines and regenerative feed heating. The thermal output of the reactor is 294 MW and the net electrical output is 92 MW, giving a net thermal efficiency of 31·6 per cent.

7.10 Fast reactors

The importance of fast reactors from the point of view of breeding has been pointed out in Chapter 3, and it has been realized from the earliest years of nuclear power that fast reactors would eventually be required if the world's resources of uranium are to be completely used for the production of energy. The world's first fast reactor was built at Los Alamos, New Mexico in 1946, and in 1951 the Experimental Breeder Reactor (EBR-1) at Arco, Idaho was the first nuclear reactor to produce electric power, though not as part of a public electricity supply system. In Britain the Dounreay Fast Reactor (60 MWth, 15 MWe) operated very successfully from 1959 to 1977, contributing a great deal of valuable information and experience, and during that time Britain had the predominant position in the world as regards fast reactor technology.

The most obvious differences between fast reactors (in which the neutron energy spectrum is in the range 10 keV to 1 MeV) and thermal reactors are that fast reactors have no moderator and require highly enriched fuel. Typically the fuel is either highly enriched uranium with 25 to 50 per cent ^{235}U, or a mixture of depleted uranium (which is nearly pure ^{238}U) and about 25 per cent ^{239}Pu. As a result, the initial fuel costs for fast reactors are very high, and to be economically viable these reactors must operate at very high ratings and burnups. For example, the rating and burnup in fast reactors are of the order of 500 MW/t and 100 000 MWd/t respectively compared with 75 MW/t and 30 000 MWd/t which are typical figures for PWRs.

The high concentration of fissile fuel and the absence of any moderator makes the core of a fast reactor very small compared with the core of a thermal reactor of the same power. For example, the British Prototype Fast Reactor (PFR) at Dounreay has a thermal power of 600 MW from a core 1 m high by 1·8 m diameter, compared with the Hunterston B reactors (AGRs) which produce 1500 MW from cores 8·2 m high by 9·3 m diameter. The power density in the PFR is almost one hundred times greater than in the AGR.

The very high ratings of fast reactors impose a number of design features which are characteristic of this type of reactor. In order to reduce the maximum fuel temperatures while at the same time achieving high coolant temperatures the fuel rods must be of very small diameter. The Dounreay PFR fuel pins contain 6 mm diameter hollow cylindrical pellets of UO_2 and PuO_2 (containing 22 to 30 per cent ^{239}Pu) enclosed in stainless steel tubes. Hexagonal fuel sub-assemblies (142 mm wide between their opposite faces) each contain 325 fuel pins held in place by spacer grids. The core consists of 70 fuel sub-assemblies.

The core is surrounded on all sides by a blanket or breeder zone in which the fuel consists of depleted uranium only. Neutrons leaking from the core into the blanket may be captured and produce ^{239}Pu, and a small fraction of the reactor power is produced in the blanket by the fission of this ^{239}Pu. Fuel sub-assemblies loaded into the core of the reactor contain depleted uranium in their lower and upper ends to form the axial blanket, with plutonium—uranium mixture in the central 914 mm section of each fuel tube to form the core. The radial blanket sub-assemblies contain depleted uranium alone. Outside them the radial reflector is made from sub-assemblies of stainless steel.

The very high rating of fast reactors makes it necessary to use a coolant with excellent heat transfer properties as well as being non-moderating. Water and heavy water are ruled out, and gases do not have adequate heat transfer properties. Liquid metals are suitable, and sodium and sodium-potassium alloy are possible coolants, the former having been used in nearly all fast reactors to date. Both have the additional advantage that their boiling points at atmospheric pressure are very high, 890°C in the case of sodium, so they do not need to be pressurized in the reactor.

The absence of highly pressurized components in the primary system of a fast reactor is a valuable safety feature. Most current designs of fast reactor are similar in that the core and primary system are in a large stainless steel vessel filled with sodium at near atmospheric pressure. An inert argon gas atmosphere above the sodium prevents any chemical reactions. Figure 7.15 shows the layout of the core, primary system and reactor vessel of this pool type of fast reactor.

The use of sodium as the coolant for a reactor has a number of disadvantages. Its melting point is 98°C, so a sodium cooled reactor must be maintained above this temperature when shut down. This problem can be overcome by the use of the sodium—potassium eutectic alloy (78 per cent K, 22 per cent Na) whose melting point is −11°C. Sodium is also highly reactive in air and water and burns readily, so high integrity of all components such as pipes and heat exchangers is essential.

Sodium also becomes activated by the ^{23}Na (n, γ)^{24}Na reaction as it passes through the reactor, and the ^{24}Na formed is radioactive. Thus the primary coolant cannot be allowed to pass outside the biological shield and a secondary sodium loop is required circulating between the primary heat

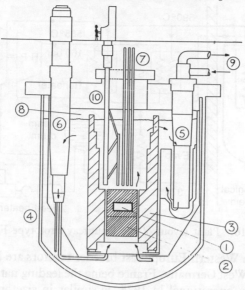

1. Core
2. Blanket and reflector
3. Neutron shield
4. Sodium filled stainless steel reactor vessel
5. Primary heat exchanger
6. Primary sodium pump
7. Control rods
8. Argon gas above sodium in the reactor vessel
9. Secondary sodium pipes to and from steam generator
10. Refuelling machine

Figure 7.15. The core and primary components of the Dounreay Prototype
Fast Reactor

exchanger which is inside the biological shield (but is separated from the
core by a neutron shield) and the steam generators.

In the Dounreay PFR, the primary radioactive sodium flows through the
hexagonal fuel sub-assemblies in the core with inlet and outlet temperatures
of 430 and 595°C respectively, and in the primary heat exchangers the
secondary sodium is heated to 590°C. In the steam generators superheated
steam is produced at 160 bar pressure and 565°C, and the electric output
is 250 MW from a core power of 600 MW. Figure 7.16 shows the layout
of the steam cycle of the Dounreay PFR.

The operation of the PFR in the years since it first went critical in 1974
has been hampered by leaks in the superheaters, evaporators and reheaters,
but in 1977 the reactor attained its full power and has since then operated
intermittently at this power. Fuel has reached nearly 20 per cent burnup,
which exceeds the objective for future commercial fast reactors of 150 000
MWd/t.

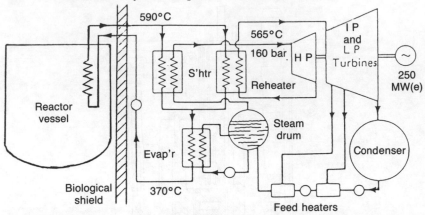

Figure 7.16.   Plant layout of the Dounreay Prototype Fast Reactor

Elsewhere in Western Europe fast breeder reactors are being developed in France and West Germany, France being the leading nation. The French Phenix reactor, operational in 1974, is similar in size and design to the Dounreay PFR. Developed from Phenix, the Superphenix 1 reactor is the first fast reactor which can accurately be described as 'commercial' in that its electrical output, 1200 MW, is comparable with other types of power stations. Superphenix 1, although primarily a French reactor, located near Lyons and supplying the French electricity grid, is the product of a consortium of French, German and Italian companies. The German SNR−300 fast breeder reactor is also the product of a consortium which includes Belgian and Dutch companies with a very small British involvement. Russia has its own fast breeder reactor, the BN−600. All these reactors, except the SNR−300, are of the pool type similar to the Dounreay PFR.

The very high costs of these reactors makes it likely that international co-operation, such as that described above, will be needed to develop commercial FBRs and their associated fuel reprocessing plant. However, a factor which at present does not encourage the development of FBRs is the abundant supply and low price of uranium for thermal reactors, a situation which provides no incentive to build FBRs. There is therefore at present some doubt about the development of FBRs in the next few decades, and even the future of the Dounreay PFR seems to be uncertain. Whereas twenty or thirty years ago it seemed certain that Britain would have commercial fast reactors by the end of this century, this is now out of the question, and it may not be until the middle of the twenty-first century that they are required. Much depends on the future price and availability of uranium for thermal reactors. In the meantime the problem exists of maintaining and improving the ability to design and build FBRs in the absence of any orders from electricity utilities.

Chapter 8

The operating characteristics
of nuclear reactors

Hitherto we have been concerned with reactors which are operating at steady state, in which the effective multiplication factor is exactly one and the neutron flux, power and temperatures are constant. These are the conditions under which power reactors operate for long periods, and for which they are designed.

It is necessary, however, to consider several other aspects of reactor operation. In the first place we will study the way in which the neutron flux varies when a reactor is non-critical. This is the state that exists when a reactor is starting up or undergoing a power increase (in which case it is supercritical), or when a reactor is shutting down or decreasing power (in which case it is subcritical). Secondly, we will study the changes which take place during prolonged operation of a reactor which have an effect on its criticality. These changes result from the burnup of fissile fuel, the production of new fissile fuel from fertile isotopes, and the buildup in a reactor of fission products with high capture cross-sections. Thirdly, we will discuss qualitatively the effects of changes of temperature within a reactor as these changes can affect the reproduction constant and introduce a feedback effect during power changes.

At all stages during the operation of a reactor a control system is essential, not only to effect startup, shutdown and power changes, but also to compensate for the changes in the reactor itself which occur during operation. Some of the principal control systems will be described.

8.1 Reactor kinetics

The study of the response of a reactor to a change in the multiplication factor from a value of one (at which the reactor is operating at steady state) to either slightly more or less than 1 is known as reactor kinetics. It is convenient at this point to define some terms of importance in reactor kinetics.

The excess reactivity δk has already been defined in Chapter 4 as:

$$\delta k = k_{\text{eff}} - 1 \qquad (8.1)$$

The reactivity ρ is defined in a slightly different way as:

$$\rho = \frac{k_{\text{eff}} - 1}{k_{\text{eff}}} \qquad (8.2)$$

or for an infinite reactor:

$$\rho = \frac{k_{\infty} - 1}{k_{\infty}} \qquad (8.3)$$

Clearly at all conditions under which reactors operate, $\rho \approx \delta k$.

Another important parameter in the study of reactor kinetics is the prompt neutron lifetime, l_{p}. In an infinite reactor the prompt neutron lifetime is the average time between the birth of prompt neutrons by fission and their final absorption in the reactor. In a thermal reactor this time is the sum of the average neutron slowing-down time (during which neutrons are slowing down from fission to thermal energy), and the average diffusion time (during which neutrons are diffusing at thermal energy up to their point of absorption). In all thermal reactors the diffusion time is much greater than the slowing-down time, typical values being about 10^{-3} seconds and 10^{-5} seconds respectively, so that the prompt neutron lifetime is very nearly equal to the diffusion time. The average diffusion time t_{d} for thermal neutrons in a reactor is:

$$t_{\text{d}} = \frac{\text{Absorption mean free path of thermal neutrons in the reactor}}{\text{Average speed of thermal neutrons}}$$

$$= \frac{\lambda_{\text{a}}}{v_{\text{av}}} \quad \text{or} \quad \frac{1}{\bar{\Sigma}_{\text{ac}} v_{\text{av}}} \qquad (8.4)$$

where $\bar{\Sigma}_{\text{ac}}$ is the average macroscopic absorption cross-section of the fuel-moderator mixture in the core of the reactor, and v_{av} is the average speed of thermal neutrons.

The diffusion time can be expressed in terms of the absorption cross-section of the moderator $\bar{\Sigma}_{\text{aM}}$ by using the equation derived in Chapter 4:

$$\bar{\Sigma}_{\text{aM}} = (1 - f)\bar{\Sigma}_{\text{ac}}$$

From this it follows that:

$$t_{\text{d}} = \frac{(1 - f)}{\bar{\Sigma}_{\text{aM}} v_{\text{av}}}$$

For a $1/v$ absorbing medium, as shown in Chapter 2:

$$\bar{\Sigma}_a v_{av} = \Sigma_{a0} v_{MP}$$

Using this result with the preceding equation, it follows that:

$$t_d = \frac{(1 - f)}{\Sigma_{a0M} v_{MP}} \tag{8.5}$$

where Σ_{a0M} is the tabulated (2200 m/s) absorption cross-section of the moderator, and $v_{MP} = 2200$ m/s. For example, the diffusion time (and also the prompt neutron lifetime) for a graphite moderated reactor in which the thermal utilization factor is 0·9 is:

$$t_d = l_p = \frac{0·1}{0·000\ 385 \times 2·2 \times 10^5}$$

$$= 1·18 \times 10^{-3} \text{ seconds}$$

This value, namely about 0·001 seconds, is typical of graphite-moderated reactors. Water-moderated, enriched uranium reactors have typical prompt neutron lifetimes of about 0·0001 seconds, and fast reactors, in which neutrons do not become thermalized, have prompt neutron lifetimes of about 10^{-7} seconds.

As was stated in Chapter 2, not all neutrons produced by fission are emitted promptly at the instant of fission. A very few neutrons, less than 1 per cent, are emitted during the radioactive decay of certain fission products, and these are called the delayed neutrons. Their properties will be described presently, however to emphasize their importance in reactor kinetics we will consider first the response of a thermal reactor to an instantaneous or step change of reactivity on the incorrect supposition that all fission neutrons are prompt. The problem may be simplified without any serious loss of accuracy by considering an infinite reactor and no neutron leakage.

The one-group diffusion equation for this problem may be written as:

$$\frac{dn}{dt} = (k_\infty - 1)\bar{\Sigma}_{ac}\phi_{th}$$

$$= (k_\infty - 1)\bar{\Sigma}_{ac} n v_{av}$$

$$= \frac{(k_\infty - 1)}{l_p} n \tag{8.6}$$

The solution of this equation for a step change of reactivity at time $t = 0$ is:

$$n = n_0 \exp\left[\frac{(k_\infty - 1)t}{l_p}\right] \tag{8.7}$$

where n_0 is the steady-state neutron density before the reactivity change. Equation (8.7) shows that the neutron density, and hence the reactor power, vary exponentially, and the rate of this change (which is an increase or decrease depending on whether the step change of reactivity is positive or negative) is characterized by the period T, which is the time for the power to change by a factor e. In terms of the period, equation (8.7) may be written as:

$$n = n_0 \, e^{t/T} \tag{8.8}$$

The period is given by:

$$T = \frac{l_p}{k_\infty - 1} \tag{8.9}$$

As an illustration of this result, consider an infinite thermal reactor in which the prompt neutron lifetime is 0·001 seconds, and k_∞ is changed from 1 to 1·001 to give an excess reactivity of 0·001. From equation (8.9) the resulting reactor period is 1 second, and in 10 seconds the reactor power increases by a factor of e^{10} or about 22 000. This is a very rapid rate of power increase which would be difficult to control. In a fast reactor a similar change in k_∞ would lead to a period of about 10^{-4} seconds, and control would be impossible. In the event of a negative step change of reactivity (for instance at shutdown), the reactor power would decrease exponentially at a rapid rate.

These results, although they have been derived for an infinite reactor, are also valid for a finite reactor, and k_∞ in the preceding equations may be replaced by k_{eff}. The effective prompt neutron lifetime in a finite reactor is however slightly less than in an infinite reactor due to neutron leakage. The reason for this is that neutrons which leak from the core have shorter than average lives. The conclusion from the preceding discussion is that, in the absence of delayed neutrons, reactor power levels would change very rapidly as a result of reactivity changes, and control during startup would be very difficult and, in the case of fast reactors, impossible.

8.2 Delayed neutrons

The delayed neutrons are produced as a result of the decay of certain radioactive fission products, mainly isotopes of bromine and iodine.

Figure 2.4 (Ch. 2) shows the decay scheme of one such isotope, ^{87}Br. The average time by which the emission of each group of neutrons is delayed is equal to the mean life of the corresponding delayed neutron precursor, which is about 80 seconds in the case of ^{87}Br. These precursors can be conveniently divided into six groups according to their half-lives, and the delayed neutron data for thermal fission in ^{235}U is shown in Table 8.1.

Table 8.1. *Delayed neutron data for thermal fission in* ^{235}U

Group	Half-life (seconds)	Mean life, t_{mi} (seconds)	Decay constant λ_i (second^{-1})	Fraction of total fission neutrons β_i
1	55·7	80·2	0·0124	0·000 215
2	22·7	32·7	0·0305	0·001 424
3	6·2	8·9	0·111	0·001 274
4	2·3	3·3	0·301	0·002 568
5	0·61	0·88	1·14	0·000 748
6	0·23	0·33	3·01	0·000 273

The total fraction of neutrons which are delayed β is:

$$\beta = \sum_{i=1}^{6} \beta_i \tag{8.10}$$

and its value for thermal fission in ^{235}U is 0·0065.

The average mean life $\bar{t}_m$ of all the delayed neutron precursors is given by the equation:

$$\bar{t}_m = \frac{1}{\beta} \sum_{i=1}^{6} \beta_i t_{mi} \tag{8.11}$$

The values of $\bar{t}_m$ and $\beta\bar{t}_m$ for ^{235}U are about 12·5 seconds and 0·08 seconds respectively. The mean lifetime l of all neutrons, both prompt and delayed is:

$$l = (1 - \beta)l_p + \beta(\bar{t}_m + l_p)$$

$$= l_p + \beta\bar{t}_m \tag{8.12}$$

Since $\beta\bar{t}_m \gg l_p$, $l \approx \beta\bar{t}_m$, which is 0·08 seconds. Returning to the earlier example and assuming that equation (8.7) can be used with a mean lifetime calculated by equation (8.12), a 0·1 per cent change of k_∞ will now result in a period of 80 seconds. This implies a very much slower rate of power rise which can be easily controlled. Unfortunately, this method does not correctly predict the response of a reactor to a step change of reactivity, and the period calculated by this method is only correct for reactivities less than about 0·0005.

8.3 Reactor kinetics with delayed neutrons

In order to correctly predict the response of a reactor to a change of reactivity it is necessary to consider the prompt neutrons, and also the production and decay of the delayed neutron precursors which produce delayed neutrons. Once again we will consider an infinite thermal reactor, using one-group theory and equation (4.40) in a modified form. The equation for the rate of change of neutron density is:

$$\begin{pmatrix} \text{The rate of change} \\ \text{of neutron density} \end{pmatrix} = \begin{pmatrix} \text{The rate of production} \\ \text{of prompt neutrons} \end{pmatrix}$$

$$+ \begin{pmatrix} \text{The rate of decay of all} \\ \text{delayed neutron precursors} \end{pmatrix}$$

$$- \begin{pmatrix} \text{The rate of} \\ \text{absorption of neutrons} \end{pmatrix}$$

$$\frac{dn}{dt} = k_\infty(1 - \beta)\Sigma_a\phi_{th} + \sum_{i=1}^{6} \lambda_i C_i - \Sigma_a\phi_{th} \qquad (8.13)$$

On the right-hand side of this equation the first term takes account of the fact that a fraction $(1 - \beta)$ of all fission neutrons are prompt. The second term is in fact six terms, one for each group of delayed neutron precursors, the symbol C_i being the concentration of the ith group of delayed neutron precursors. $\lambda_i C_i$ is the rate of decay of the ith precursor and the rate of production of delayed neutrons from this particular source. Equation (8.13) can be rewritten as:

$$\frac{dn}{dt} = \frac{k_\infty(\rho - \beta)}{l_p} n + \sum_{i=1}^{6} \lambda_i C_i \qquad (8.14)$$

There are six equations for the concentrations of the six groups of delayed neutron precursors, each one of the form:

$$\begin{pmatrix} \text{The rate of change of the} \\ \text{concentration of the } i\text{th group} \\ \text{of delayed neutron precursors} \end{pmatrix} = \begin{pmatrix} \text{The rate of formation of} \\ \text{the } i\text{th group of precursors} \end{pmatrix}$$

$$- \begin{pmatrix} \text{The rate of decay of the} \\ i\text{th group of precursors} \end{pmatrix}$$

$$\frac{dC_i}{dt} = k_\infty \beta_i \bar{\Sigma}_a n v_{av} - \lambda_i C_i$$

$$= \frac{k_\infty \beta_i n}{l_p} - \lambda_i C_i \quad (i = 1 \text{ to } 6) \tag{8.15}$$

(six equations)

The response of a reactor to a step change of reactivity is determined by solving the seven simultaneous linear differential equations (8.14) and (8.15). The solution is laborious and the result can most easily be obtained by solving the equations on an analogue or digital computer.

The essential features of the solution can be seen by simplifying the problem and replacing the six groups of delayed neutron precursors by a single group. The value of β for the single group of delayed neutrons is given by equation (8.10), and the value of the decay constant λ of the single group of delayed neutron precursors is given by:

$$\lambda = \frac{1}{\bar{t}_m}$$

where $\bar{t}_m$ is given by equation (8.11). The value of λ for a ^{235}U fuelled reactor is about $0 \cdot 08$ second^{-1}.

The seven equations (8.14) and (8.15) can now be reduced to two equations:

$$\frac{dn}{dt} = \frac{k_\infty(\rho - \beta)}{l_p} n + \lambda C \tag{8.16}$$

and:

$$\frac{dC}{dt} = \frac{k_\infty \beta}{l_p} n - \lambda C \tag{8.17}$$

where C is now the concentration of all the delayed neutron precursors. By inspecting equation (8.16) it is possible to get a qualitative idea of the response of a reactor to a positive step change of reactivity. If ρ is positive and less than β, the first term on the right-hand side of equation (8.16) is negative, and the increase of the neutron density is governed by the term λC, the rate at which the delayed neutron precursors decay, which as we have seen is quite slow. If $\rho = \beta$, then $dn/dt = \lambda C$, the reactor is critical on prompt neutrons alone and the rate of increase of the neutron density is equal to the rate of production of delayed neutrons. In this condition the reactor is referred to as being prompt critical. If ρ is greater than β, then both terms on the right-hand side of equation (8.16) are positive and the first term produces an exponential increase with a short period because of the small value of l_p. This is a similar situation to that which existed in the case in which delayed neutrons were neglected. The conclusion to be drawn from this argument is that provided ρ is less than β, the rate of power increase following a positive step change of reactivity is fairly slow and can be controlled. On the other hand if ρ is greater than β the rate of power increase is rapid and difficult (if not impossible) to control. The criterion for safe operation during startup and power increase is that the reactor must not be allowed to become prompt critical.

Equations (8.16) and (8.17) may be solved by standard methods. If solutions of the form:

$$n = \sum A \, e^{\omega t}$$

and

$$C = \sum B \, e^{\omega t}$$

are assumed, and these are substituted into equations (8.16) and (8.17), and the approximation $k_\infty = 1$ is made, the following quadratic equation for ω is obtained:

$$\omega^2 l_p + \omega(\beta - \rho + l_p \lambda) - \lambda \rho = 0$$

If $(\beta - \rho + l_p \lambda)^2$ is much greater than $|2l_p \lambda \rho|$, and if $l_p \lambda$ is much less than $(\beta - \rho)$, the two values of ω satisfying the above equation are approximately:

$$\omega_1 \approx \frac{\lambda \rho}{(\beta - \rho)} \quad \text{and} \quad \omega_2 \approx \frac{-(\beta - \rho)}{l_p}$$

The validity of these approximations depends on the value of ρ being less than about $\beta/2$, as can be verified by substituting the following values into the above expressions:

$$\beta = 0.0065, \quad \rho = 0.003, \quad \lambda = 0.08 \text{ s}^{-1}, \quad l_p = 0.001 \text{ s}$$

The complete solution for n has two terms:

$$n = A_1 \, e^{\omega_1 t} + A_2 \, e^{\omega_2 t}$$

The values of A_1 and A_2 can be found to be given approximately by:

$$A_1 \approx \frac{\beta}{\beta - \rho} n_0 \quad \text{and} \quad A_2 \approx \frac{-\rho}{\beta - \rho} n_0$$

where n_0 is the steady-state neutron density prior to the step change of reactivity. The complete solution for n is:

$$n = n_0 \left\{ \frac{\beta}{\beta - \rho} \exp\left(\frac{\lambda \rho}{\beta - \rho} t \right) - \frac{\rho}{\beta - \rho} \exp\left(-\frac{\beta - \rho}{l_p} t \right) \right\}$$

$$(8.18)$$

This equation is valid, as already noted, if ρ is less than about $\beta/2$. If ρ is positive the solution consists of a positive term with a positive exponent whose period is $(\beta - \rho)/\lambda\rho$, and a negative term with a negative exponent which decays rapidly as its period is $l_p/(\beta - \rho)$. The complete solution for a ^{235}U fuelled thermal reactor with a prompt neutron lifetime of 0·001 second following a step change of reactivity of +0·002 is shown in Figure 8.1.

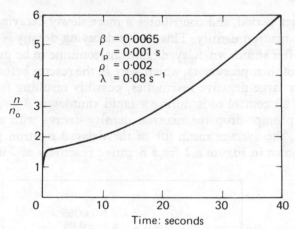

$$\beta = 0\cdot0065$$
$$l_p = 0\cdot001 \text{ s}$$
$$\rho = 0\cdot002$$
$$\lambda = 0\cdot08 \text{ s}^{-1}$$

Time: seconds

Figure 8.1.   The response of a reactor to a step change of reactivity

From Figure 8.1 it can be seen that after an initial rapid rise which causes the neutron density and reactor power to increase by a factor $\beta/(\beta - \rho)$ in about 1 second, the rate of increase is slowed down and the neutron density increases exponentially with a stable period.

$$T = \frac{\beta - \rho}{\lambda\rho} \tag{8.19}$$

For example, a positive step change of reactivity of 0·002 causes the power to increase promptly by a factor 1·44, and thereafter to increase exponentially with a stable period of about 28·1 seconds, which makes the reactor easily controllable. Smaller reactivities result in a smaller prompt increase and a longer stable period.

While these figures are not exact due to the inherent inaccuracy of the 'one group of delayed neutrons' approximation, they do give a qualitatively correct idea of the way in which a reactor responds to an increase of reactivity, and for small reactivities (ρ less than about 0·1 β) the results are quite accurate.

The response of a reactor to a negative step change of reactivity, as would occur at shutdown, can also be deduced from equation (8.18). In this case ρ is negative and both parts of the solution for n have positive coefficients and negative exponents. The term

$$\frac{\rho}{\beta - \rho} \exp\left(-\frac{\beta - \rho}{l_{\mathrm{p}}} t\right)$$

has a very short period and therefore contributes a rapid drop in the neutron density. The term

$$\frac{\beta}{\beta - \rho} \exp\left(-\frac{\lambda\rho}{\beta - \rho} t\right)$$

has a longer period, and contributes a more slowly decaying component to the neutron density. This slowly decaying density is due to the fact that after shutdown delayed neutrons continue to be produced by the decay of their precursors, which were in the reactor before the shutdown. For large negative reactivities, possibly resulting from the insertion of all control rods during a rapid shutdown, $\beta - \rho \approx \rho$, and after the prompt drop the neutron density decays with a period of about $1/\lambda$, the average mean life of the delayed neutron precursors. This is shown in Figure 8.2 for a negative reactivity of 0·05.

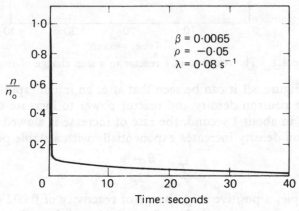

Figure 8.2.   The response of a reactor to a negative step change of reactivity

The 'one group of delayed neutrons' approximation does not accurately predict the final rate of decrease of the neutron density after shutdown. According to the approximate solution the period is $1/\lambda$, which is about 12 seconds for a ^{235}U fuelled reactor. In fact the neutron density finally decays with a period equal to the mean life of the longest lived delayed neutron precursor, which is about 80 seconds. This means that it is impossible to shut down a reactor completely in a matter of seconds, and many minutes may elapse before the power of a reactor operating at several megawatts drops to a few watts. This effect is even more pronounced in reactors containing large amounts of deuterium (in heavy water) or beryllium. Gamma radiation from radioactive fission products induces (γ, n) reactions in these elements and produces neutrons long after shutdown. Consequently, the time for the reactor power to fall to a very low level is prolonged. A similar effect occurs as a result of the heat released by the decay of radioactive fission products, and this will be described in Chapter 10.

8.4 Reactivity changes in an operating reactor

Burnup of fuel

During reactor operation, fissile material is being consumed or burned up at a rate proportional to the power of the reactor. In a reactor containing fertile material, new fissile material is also being created by one of the processes described in Chapter 3. It is necessary in the design of a reactor to be able to estimate the fuel concentration as a function of time to ensure that the reactor has sufficient fuel to enable it to remain critical for a specified time, possibly a year or two. The accurate prediction of the fuel concentration in a reactor over a prolonged period is difficult, and we will restrict outselves to a qualitative discussion.

In an infinite reactor fuelled with pure ^{235}U, operating at constant power, the rate of fission and hence the rate of burnup is constant and uniform throughout the reactor, and the ^{235}U concentration decreases linearly with time and uniformly throughout the reactor. In a finite reactor fuelled with ^{235}U, the rate of burnup is greatest at the centre of the core where the flux is greatest, thus the composition of the reactor varies non-uniformly, the flux shape changes, and the problem becomes more complicated.

The important feature of a reactor fuelled with highly enriched uranium or pure ^{235}U is that the fuel concentration, and hence the excess reactivity of the reactor, decrease continuously as operation proceeds.

In a reactor containing a large amount of fertile material, for example a natural or slightly enriched uranium reactor, the burnup of the original fissile isotope is offset to some extent by the production of new fissile material. In the case of a uranium fuelled reactor the important processes are:

$$^{235}U + n \left\{ \begin{array}{l} ^{236}U \\ \text{fission} \end{array} \right.$$

$$^{238}U + n \rightarrow {}^{239}U \rightarrow {}^{239}Np \rightarrow {}^{239}Pu$$

$$^{239}Pu + n \left\{ \begin{array}{l} ^{240}Pu \\ \text{fission} \end{array} \right.$$

$$^{240}Pu + n \rightarrow {}^{241}Pu$$

$$^{241}Pu + n \left\{ \begin{array}{l} ^{242}Pu \\ \text{fission} \end{array} \right.$$

The creation of fissile ^{239}Pu is important during the early stages of reactor operation when burnup is low; however, ^{241}Pu becomes significant at high burnup. The equations for the concentrations of ^{235}U, ^{238}U and ^{239}Pu are:

^{235}U: $$\frac{dN(^{235}U)}{dt} = -\phi\sigma_a(^{235}U)N(^{235}U)$$

^{238}U: $$\frac{dN(^{238}U)}{dt} = -\phi\sigma_c(^{238}U)N(^{238}U)$$

^{239}Pu: $$\frac{dN(^{239}Pu)}{dt} = \phi\sigma_c(^{238}U)N(^{238}U) - \phi\sigma_a(^{239}Pu)N(^{239}Pu)$$

(In these equations the terms on the right-hand sides represent total rates of neutron absorption at all neutron energies, and include resonance capture in ^{238}U.)

To simplify the solution, which would otherwise be rather complicated, we will consider an infinite reactor operating at constant flux, and assume that the concentration of ^{238}U is constant. (The value of $\phi\sigma_c$ (^{238}U) is much less than either $\phi\sigma_a$ (^{235}U) or $\phi\sigma_a$ (^{239}Pu).)

The solutions of the above equations for $N(^{235}U)$ and $N(^{239}Pu)$ are:

$$N(^{235}U) = N_0(^{235}U) \exp\left[-\phi\sigma_a(^{235}U)t\right] \qquad (8.20)$$

where $N_0(^{235}U)$ is the initial concentration of ^{235}U, and:

$$N(^{239}Pu) = N_0(^{238}U)\frac{\phi\sigma_c(^{238}U)}{\phi\sigma_a(^{239}Pu)}\left\{1 - \exp\left[-\phi\sigma_a(^{239}Pu)t\right]\right\} \quad (8.21)$$

where $N_0(^{238}U)$ is the initial concentration of ^{238}U.

Figure 8.3 shows approximately the variation of the ^{235}U and ^{239}Pu concentrations in a natural uranium fuelled reactor operating at constant flux. In the early stages of operation the production of ^{239}Pu more than offsets the burnup of ^{235}U, and there is an increase of reactivity. In due course this trend is reversed, the concentration of

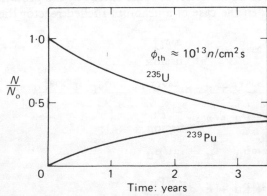

Figure 8.3. The concentration of ^{235}U and ^{239}Pu in an operating reactor

^{239}Pu tends to become constant while the concentration of ^{235}U continues to decrease. Consequently the reactivity decreases and after some time the reactor will become subcritical and require refuelling.

8.5 Reactor poisons

Of the many isotopes that are formed as fission products in a reactor, two are of particular importance because they have very high absorption cross-sections, and their presence in the reactor has a considerable effect on reactivity. These two isotopes are xenon 135 and samarium 149, and they are known as reactor poisons.

Xenon poisoning

Xenon 135 is the most important fission product poison. Its average thermal neutron absorption cross-section at 20°C is 2.75×10^6 barns, and it is formed directly as a fission product and also in the decay chain of the fission product tellurium 135. The tellurium decay chain is

$$^{135}\text{Te} \xrightarrow[T_{1/2}<1\text{ min}]{} {}^{135}\text{I} \xrightarrow[T_{1/2}=6\cdot7\text{ h}]{} {}^{135}\text{Xe} \xrightarrow[T_{1/2}=9\cdot2\text{ h}]{} {}^{135}\text{Cs} \rightarrow {}^{135}\text{Ba (stable)}$$

and the fission yield of ^{135}Te from ^{235}U fission is 0·061 atoms per fission. The half-life of ^{135}Te is so short, however, that ^{135}I may be considered the primary fission product. Xenon 135 is also produced as a primary fission product to the extent of 0·003 atoms per fission.

The effect of ^{135}Xe on the reactivity of a reactor depends on its concentration during steady power operation and after shutdown. The equation for the concentration $N(\text{I})$ of ^{135}I from which it is formed can be written as:

$$\begin{pmatrix} \text{The rate of change} \\ \text{of } {}^{135}\text{I concentration} \end{pmatrix} = \begin{pmatrix} \text{The rate of} \\ \text{formation of } {}^{135}\text{I} \end{pmatrix} - \begin{pmatrix} \text{The rate of} \\ \text{decay of } {}^{135}\text{I} \end{pmatrix}$$

$$\frac{dN(\text{I})}{dt} = \gamma(\text{I})\bar{\Sigma}_f\phi_{\text{th}} - \lambda(\text{I})N(\text{I}) \tag{8.22}$$

where $\gamma(\text{I})$ is the fission yield of ^{135}I. The solution of this equation is:

$$N(\text{I}) = \frac{\gamma(\text{I})\bar{\Sigma}_f\phi_{\text{th}}}{\lambda(\text{I})} (1 - e^{-\lambda(\text{I})t}) \tag{8.23}$$

where t is the time after startup of an initially clean reactor (i.e. no fission products). The ^{135}I concentration reaches equilibrium after about 30 hours operation, the value being:

$$N(\text{I})_{\text{eq}} = \frac{\gamma(\text{I})\bar{\Sigma}_f\phi_{\text{th}}}{\lambda(\text{I})} \tag{8.24}$$

The equation for the ^{135}Xe concentration $N(\text{Xe})$ is:

$$\begin{pmatrix}\text{The rate of change of}\\ {}^{135}\text{Xe concentration}\end{pmatrix} = \begin{pmatrix}\text{The rate of production}\\ \text{of } {}^{135}\text{Xe by decay of } {}^{135}\text{I}\end{pmatrix}$$

$$+ \begin{pmatrix}\text{The rate of production of}\\ {}^{135}\text{Xe as a fission product}\end{pmatrix}$$

$$- \begin{pmatrix}\text{The rate of}\\ \text{decay of } {}^{135}\text{Xe}\end{pmatrix}$$

$$- \begin{pmatrix}\text{The rate of neutron}\\ \text{capture in } {}^{135}\text{Xe}\end{pmatrix}$$

$$\frac{dN(\text{Xe})}{dt} = \lambda(\text{I})N(\text{I}) + \gamma(\text{Xe})\bar{\Sigma}_f\phi_{\text{th}} - \lambda(\text{Xe})N(\text{Xe}) - \bar{\sigma}_c(\text{Xe})N(\text{Xe})\phi_{\text{th}}$$

$$(8.25)$$

where $\gamma(\text{Xe})$ is the fission yield of ^{135}Xe. Like the ^{135}I, the ^{135}Xe reaches an equilibrium concentration after about two days' operation, the value being found from equations (8.24) and (8.25) to be:

$$N(\text{Xe})_{\text{eq}} = \frac{\{\gamma(\text{I}) + \gamma(\text{Xe})\}\bar{\Sigma}_f\phi_{\text{th}}}{\lambda(\text{Xe}) + \bar{\sigma}_c(\text{Xe})\phi_{\text{th}}} \qquad (8.26)$$

The effect of this equilibrium concentration of ^{135}Xe on the reactivity of a reactor can be determined. An infinite homogeneous reactor will be considered, and it will be assumed that the presence of ^{135}Xe affects only the value of the thermal utilization factor. The change of reactivity may be written as

$$\rho(\text{Xe}) = \frac{k'_\infty - k_\infty}{k'_\infty} = \frac{f' - f}{f'} = f\left(\frac{1}{f} - \frac{1}{f'}\right)$$

where k'_∞ and f' refer to the poisoned reactor. Expressions for f' and f are:

$$f' = \frac{\bar{\Sigma}_{\text{aF}}}{\bar{\Sigma}_{\text{aF}} + \bar{\Sigma}_{\text{cM}} + \bar{\Sigma}_c(\text{Xe})}$$

and

$$f = \frac{\bar{\Sigma}_{\text{aF}}}{\bar{\Sigma}_{\text{aF}} + \bar{\Sigma}_{\text{cM}}}$$

Consequently

$$\frac{1}{f} - \frac{1}{f'} = -\frac{\bar{\Sigma}_c(\text{Xe})}{\bar{\Sigma}_{\text{aF}}} = -\frac{N(\text{Xe})\bar{\sigma}_c(\text{Xe})}{\bar{\Sigma}_{\text{aF}}}$$

and
$$\rho(\text{Xe}) = -f \, \frac{N(\text{Xe})\bar{\sigma}_c(\text{Xe})}{\bar{\Sigma}_{aF}}$$

Using equation (8.26) for $N(\text{Xe})_{eq}$, and noting that $\bar{\Sigma}_f/\bar{\Sigma}_{aF} = \eta/\nu$,

$$\rho(\text{Xe}) = -\frac{f\eta}{\nu} \, \frac{\{\gamma(\text{I}) + \gamma(\text{Xe})\}}{\left(\dfrac{\lambda(\text{Xe})}{\bar{\sigma}_c(\text{Xe})\phi_{th}} + 1\right)} \tag{8.27}$$

Evidently $\rho(\text{Xe})$ increases as ϕ_{th} increases, and reaches a limiting value when $\phi_{th} \gg \lambda(\text{Xe})/\bar{\sigma}_c(\text{Xe})$. The value of this ratio is about 0.75×10^{13} cm^{-2} s^{-1}, so when ϕ_{th} is 10^{14} neutrons/cm^2 s, a typical value for a power reactor, this limiting value is nearly reached. It is:

$$\rho(\text{Xe}) \text{ maximum} = -\frac{f\eta\{\gamma(\text{I}) + \gamma(\text{Xe})\}}{\nu} \tag{8.28}$$

Consider a natural uranium fuelled reactor in which $\eta = 1.32, \nu = 2.42$ and $\{\gamma(\text{I}) + \gamma(\text{Xe})\} = 0.064$. If we assume that $f = 0.9$, then the maximum value of $\rho(\text{Xe})$ is -3.1 per cent. This figure represents the amount of excess reactivity which must be built into the reactor to enable the maximum possible effect of ^{135}Xe poisoning to be overcome during steady-state operation.

It is also important to consider the way in which the ^{135}Xe concentration varies after a rapid reactor shutdown. (The ^{135}Xe may be assumed to be at its equilibrium concentration before shutdown.) Immediately after shutdown the rate of production of ^{135}Xe decreases to a fraction $\gamma(\text{I})/[\gamma(\text{I}) + \gamma(\text{Xe})]$ of the pre-shutdown value, while the rate of elimination decreases to a fraction $\lambda(\text{Xe})/[\lambda(\text{Xe}) + \bar{\sigma}_c(\text{Xe})\phi_{th}]$ of the pre-shutdown value. In a power reactor, whose average thermal neutron flux is 10^{12} neutrons/cm^2 s or more, the second fraction is smaller than the first, and the ^{135}Xe concentration rises after shutdown. The rate of rise depends on the pre-shutdown flux, and increases as this flux increases. In due course, as the ^{135}I decays, the ^{135}Xe concentration reaches a peak (which also increases as the pre-shutdown flux increases) and thereafter decays. This behaviour is shown in Figure 8.4.

The important point concerning the peak ^{135}Xe concentration after shutdown is that unless sufficient excess reactivity is built into the reactor to overcome the increased poisoning effect, it may not be possible to restart the reactor until many hours have elapsed. To override the maximum xenon poisoning in a reactor operating at 10^{14} n/cm^2 s an excess reactivity of about 13 per cent is required. If less reactivity is available, for example 6 per cent (*see* Figure 8.4), then after shutdown

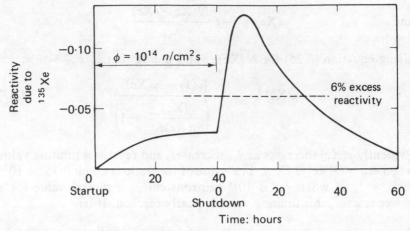

Figure 8.4.   Xenon 135 concentration in a reactor during operation and after shutdown

about 2 hours are available to restart the reactor before the ^{135}Xe builds up, and failing this it is necessary to wait for about 28 hours until the ^{135}Xe has passed its maximum and decayed sufficiently.

Samarium poisoning

Samarium 149 is formed from the fission product neodymium 149 by the decay chain

$$^{149}\text{Nd} \xrightarrow[T_{1/2}=2\text{h}]{} {}^{149}\text{Pm} \xrightarrow[T_{1/2}=54\text{h}]{} {}^{149}\text{Sm (stable)}$$

The half-life of ^{149}Nd is short compared with that of the intermediate product promethium 149, and the latter may be considered as the primary fission product. The fission yield of ^{149}Pm, $\gamma(\text{Pm})$, is 0·0113 atoms per fission, and the average thermal neutron capture cross-section of ^{149}Sm is $5\cdot8 \times 10^4$ barns.

The concentration of ^{149}Sm may be found from the equations:

$$\frac{\mathrm{d}N(\text{Pm})}{\mathrm{d}t} = \gamma(\text{Pm})\overline{\Sigma}_\text{f}\phi_\text{th} - \lambda(\text{Pm})N(\text{Pm}) \tag{8.29}$$

and

$$\frac{\mathrm{d}N(\text{Sm})}{\mathrm{d}t} = \lambda(\text{Pm})N(\text{Pm}) - N(\text{Sm})\bar{\sigma}_\text{c}(\text{Sm})\phi_\text{th} \tag{8.30}$$

where $N(\text{Pm})$ and $N(\text{Sm})$ are the concentrations of ^{149}Pm and ^{149}Sm respectively. The equilibrium concentration of ^{149}Sm, which is reached after several days operation, is:

$$N(\text{Sm})_\text{eq} = \frac{\gamma(\text{Pm})\overline{\Sigma}_\text{f}}{\bar{\sigma}_\text{c}(\text{Sm})} \tag{8.31}$$

This concentration is independent of the reactor flux or power, and its effect on the reactivity is:

$$\rho(\text{Sm}) = -\frac{f\eta\gamma(\text{Pm})}{\nu} \qquad (8.32)$$

From the figures used previously for a natural uranium fuelled reactor, the value of $\rho(\text{Sm})$ is -0.55 per cent.

After shutdown the ^{149}Sm in the reactor, being stable, remains, while the ^{149}Pm in the reactor decays to produce more ^{149}Sm, whose concentration therefore rises to a higher level. The post-shutdown equilibrium concentration is:

$$N(\text{Sm})_{\text{eq}} \text{ after shutdown} = \frac{\gamma(\text{Pm})\overline{\Sigma}_{\text{f}}}{\overline{\sigma}_{\text{c}}(\text{Sm})} \left(1 + \frac{\overline{\sigma}_{\text{c}}(\text{Sm})\phi_{\text{th}}}{\lambda(\text{Pm})}\right) \qquad (8.33)$$

The variation of ^{149}Sm in an operating reactor is shown in Figure 8.5.

In a reactor operating at a thermal neutron flux of 10^{14} n/cm^2 s the reactivity effect of ^{149}Sm after shutdown rises to a constant value of -1.45 per cent. and there must be sufficient excess reactivity available

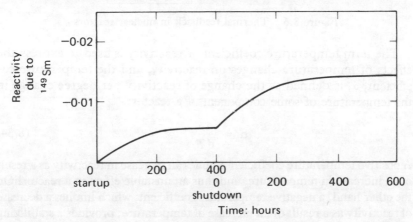

Figure 8.5. Samarium 149 concentration in a reactor during operation and after shutdown

in the reactor to overcome this. It is obvious, however, from the figures that have been calculated for ^{135}Xe and ^{149}Sm that the former is much more important as a reactor poison.

8.6 Temperature effects on reactivity

Changes in the temperature of a reactor which result from a change of power can cause changes in reactivity which in turn affect the power. Thus a feedback process is established whose characteristics have an important bearing on the safety of the reactor. For instance, if an increase in power followed by an increase in the temperature of parts of a reactor causes an increase in reactivity, this will lead to a further increase in power and an unstable situation will exist which, if it is not controlled, could lead to an accident. On the other hand if an increase in power and temperature leads to a decrease in reactivity, the original power rise will be retarded or reversed and a stable situation will exist in which the reactor tends to control itself. This is obviously desirable from the point of view of safety (see Figure 8.6).

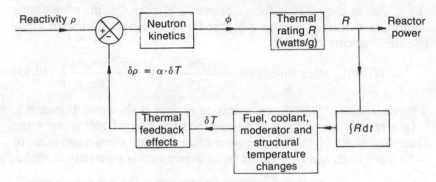

Figure 8.6.   Thermal feedback in nuclear reactors

The term temperature coefficient of reactivity is used to express the effects of temperature changes on reactivity, and the temperature coefficient, α, is defined as the change of reactivity per degree change in the temperature of some component of a reactor:

$$\alpha = \frac{d\rho}{dT} \qquad (8.34)$$

A positive temperature coefficient implies an increase in reactivity as a result of an increase in temperature, and thus an unstable effect in a reactor. On the other hand, a negative temperature coefficient, which implies a decrease of reactivity as a result of an increase of temperature, provides a stabilizing effect.

When a reactor undergoes a power increase, the resulting temperature changes in different components of the reactor occur at different rates. The fuel temperature will rise at nearly the same rate as the power with little or no lag. The coolant temperature will rise more slowly because of the time lag in the transfer of heat from the fuel to the coolant. Finally (in the case of a large graphite-moderated reactor), the moderator temperature will rise much more slowly due to its large mass and thermal capacity. The temperature coefficients of the fuel, coolant and moderator must be considered separately, and it must be borne in mind that they act at different rates.

It is particularly important that the fuel temperature coefficient, which acts with little or no delay in a power rise, should be negative for safe operation. All existing types of power reactors do have negative fuel temperature coefficients due to the phenomenon known as 'Doppler broadening of the resonances'. This effect is due to the fact that as the temperature of the fuel rises, the thermal vibration of the fuel nuclei also increases. Consequently the range of neutron energies which corresponds to the increased thermal vibration of the fuel nuclei also increases, and the resonance peaks in the absorption cross-sections of the fuel nuclei (refer to section 2.9 and Figures 2.9 and 2.12) are broadened. See Figure 8.7 for an illustration of this effect. Overall, the increase in temperature and broadening of the resonances leads to increased resonance neutron absorption in the fuel, and in a reactor in which the most abundant fuel isotope is ^{238}U (and this applies to all existing types) the increased neutron capture in ^{238}U is the most important effect. Thus the Doppler coefficient of reactivity is negative in all existing types of reactor. Its value is typically of the order of $-10^{-5}\ \delta k/k$.

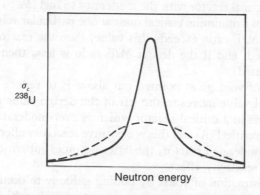

Figure 8.7. Illustration of the effect of Doppler broadening of a resonance in the capture cross-section of ^{238}U. Solid line shows the shape of the resonance at low temperature. Dotted line shows the shape at elevated temperature

This effect is of fundamental importance because the ^{238}U, which is responsible for the Doppler effect, is an integral part of the reactor fuel, being mixed with the ^{235}U or (in the case of fast reactors) the ^{239}Pu in which energy is being released by fission. Any rise in the fuel temperature due to an increased fission rate causes a decrease of reactivity due to the Doppler effect with no time delay at all.

In light water-cooled and moderated reactors, the coolant temperature affects the reactivity primarily as a result of any nucleate or bulk boiling that may occur. The production of steam bubbles or voids by boiling reduces the average density of the water in the core and also reduces the moderator to fuel ratio. This may affect the reactivity in the following ways:

1. Reduction of the average density of the moderator causes neutron mean free paths in the core to be increased, thus increasing neutron leakage. This is a negative reactivity effect.

2. Reduction in the moderator to fuel ratio increases the fraction of neutrons captured in ^{238}U resonances during slowing down. This is also a negative reactivity effect.
3. Reduction in the moderator to fuel ratio reduces the fraction of thermal neutrons captured in the moderator and the thermal utilization factor increases. This is a positive reactivity effect.
4. Hardening of the thermal neutron spectrum due to decreased moderation of neutrons increases the average energy of the thermal neutrons. At this slightly higher energy the fraction of neutrons absorbed in ^{235}U decreases due to the decrease in the value of $\sigma_a(^{235}$U$)$ relative to $\sigma_c(^{238}$U$)$. This is a negative reactivity effect.

The overall void coefficient of reactivity is the sum of these effects, and in general depends upon whether the reactor is designed to be over- or under-moderated. Figure 8.8 shows the typical variation of the critical mass of fuel in a thermal reactor with the moderator to fuel (M/F) ratio, showing the existence of a minimum critical mass at one particular value of this ratio. If the design M/F ratio exceeds this value, then the reactor is said to be over-moderated, and if the design M/F ratio is less, then the reactor is under-moderated.

The effect of voidage as pointed out above is to reduce the M/F ratio and the actual value moves to the left of the design value on Figure 8.8. If this happens in a critical reactor which is over-moderated, the reactor becomes supercritical (A) and this is a positive reactivity effect. If it happens in an under-moderated reactor, the latter becomes subcritical (B) and this is a negative reactivity effect.

Since the formation of voids by boiling is likely to occur very quickly following an increase in reactor power, a negative void coefficient is desirable

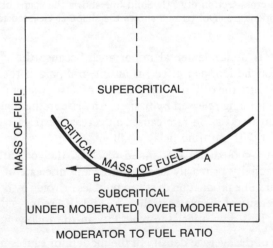

Figure 8.8.   The variation of critical fuel mass with moderate to fuel ratio, showing the effect of voidage on reactivity

from the viewpoint of reactor safety. This point will be considered further in the description in Chapter 11 of the accident to the Russian reactor at Chernobyl in 1986.

In reactors other than PWRs and BWRs the functions of the moderator and coolant are carried out by different substances. For example an AGR has graphite moderator and carbon dioxide coolant. Even in the CANDU design, in which both cooling and moderation are done by heavy water, the D_2O coolant is physically separated from the D_2O moderator, and they are at different temperatures during reactor operation. Thus in such reactors it is necessary to consider the moderator temperature coefficient as distinct from the coolant temperature (or void) coefficient.

As the moderator temperature rises, the average energy of the thermal neutrons (which are in energy equilibrium with the moderator) also rises. The fission cross-section of ^{235}U generally decreases with increasing neutron energy in the energy range $0 \cdot 01$ to $0 \cdot 1$ eV, the upper part of the thermal neutron spectrum (see Figure 2.12), so a rise in moderator temperature leads to a reduction in fission in ^{235}U relative to capture in ^{238}U, thus reducing the factor eta (η) and the reactivity ρ. In this respect the moderator temperature coefficient is negative.

The moderator temperature coefficient can become positive if the fuel has been in the reactor for a long time and a significant amount of ^{239}Pu has been produced by neutron capture in ^{238}U (see Figure 8.3). This isotope has a pronounced resonance in its absorption cross-section at $0 \cdot 3$ eV, and an increase in the moderator temperature, increasing the average energy of thermal neutrons towards this resonance, results in an increase in the rate of fission in ^{239}Pu. This is a positive reactivity effect which is particularly evident in natural uranium fuelled reactors, e.g. Magnox and CANDU designs. However, in these two reactor types the moderator has a large mass and is physically separated from the coolant, so the temperature of the moderator rises quite slowly in any power increase, no matter how rapid the latter is. Thus the time constant for the moderator temperature coefficient is long, and even if the coefficient is positive, it is readily controllable.

Finally, a rise in temperature of the reactor structure causes thermal expansion of the structure. This expansion implies that the mean inter-atomic distance in the reactor structure has increased. Hence the mean free path of neutrons is increased, and the probability of a neutron leaking out of the core (instead of being reflected back into the core) is increased. Similarly thermal expansion of the control rods means that a larger portion of the rods are in the core, and thus that neutron capture within them (a form of leakage) is more likely. Hence the reactivity coefficient due to structural temperature is normally negative in well-designed reactors. This effect, like the moderator temperature coefficient, is normally much slower to act than the fuel temperature or void coefficients.

The effects of the various reactivity coefficients are summarized in Figure 8.9.

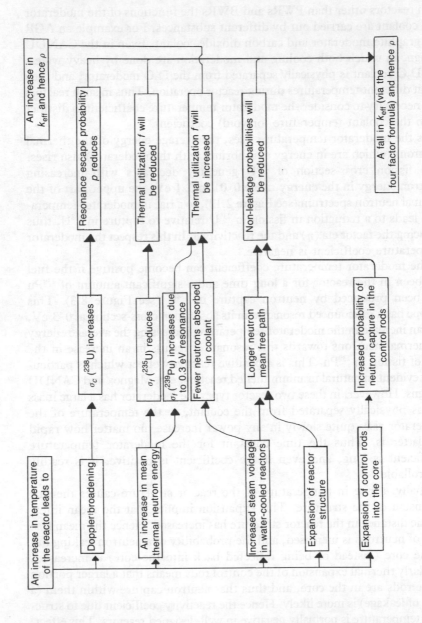

Figure 8.9. The ways in which temperature affects reactivity

An example of the way in which reactor power responds to a change in reactivity is given in Figure 8.10, which shows the response of a heavy water moderated and cooled reactor to a change in reactivity caused by withdrawal of a control rod ($\rho = 0 \cdot 003$ at the rate of $0 \cdot 000\ 25$ per second). The fuel and moderator temperature coefficients are both negative, and after an initial power and temperature rise, the reactor stabilizes itself at a higher power without further control rod movements. This is typical of most power reactors over their entire power ranges.

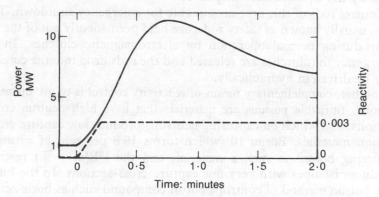

Figure 8.10. The response of a reactor with negative temperature coefficient to an increase of reactivity

8.7 Control systems in reactors

The control system of a reactor is required to perform three functions, namely:

1. Bring about the small changes of reactivity necessary to start up the reactor, change its power level as desired, and shut down the reactor.
2. Absorb the built-in excess reactivity, and compensate for the spontaneous reactivity changes due to fuel burnup, fission product poisoning and temperature effects.
3. Provide a means of shutting down the reactor rapidly in an emergency by inserting a large amount of negative reactivity.

In many reactors two or three control systems may be used, one for each of these functions; however, different reactor types employ different systems. Two more considerations are that the control system should not absorb neutrons wastefully, and it should not distort the flux in the reactor more than necessary.

The most common control system for thermal reactors consists of rods containing a neutron-absorbing material such as boron, cadmium or hafnium. These rods are moved into and out of the core by drive mechanisms which can be precisely controlled. This system is good for startup, power changes and shutdown, but is less suitable for absorbing all the reactor's built-in reactivity. This is because at the start of the reactor's life some or all the control rods would be partially in the core for a long period, neutrons would be wastefully absorbed and the flux in the core distorted with the result that fuel temperatures would differ from, and in some cases be greater than, the values calculated by the equations of Chapter 6.

Control rods of this type are suitable for emergency shutdown. The rods, usually known as safety rods, are held permanently out of the reactor during normal operation by electromagnetic clutches. In an emergency the clutches are released and the rods drop into the core, or may be driven in hydraulically.

Another, complementary means of reactivity control is to use burnable poisons. Burnable poisons are materials that have high neutron cross-sections, and which on capturing neutrons produce low capture cross-section materials. Boron 10 (which forms 18·8 per cent of naturally occurring boron) is such a material, and the $^{10}B(n, \alpha)$ ^7Li reaction produces isotopes with very low capture cross-sections. In the burnable poison method of control a boron compound such as boric acid is mixed with the moderator to absorb the excess reactivity of a new reactor. As operation proceeds, the ^{10}B is 'burned out' by neutron capture and its negative reactivity effect is reduced, thus compensating for fuel burnup. The concentration of boron in the moderator can be controlled either by injecting more boron or by circulating the moderator through an ion exchange column to remove boron. This method has the advantage that the flux is not perturbed by the presence of boric acid throughout the core, however it cannot be used for controlling startup, power variation and xenon fluctuations so control rods are still necessary. Burnable poisons are used in water-cooled reactors which have off-load refuelling, since they enable longer periods of power generation (in between refuelling shutdowns) to be achieved.

Burnable poisons are also used in Britain's Advanced Gas-cooled Reactors (AGRs) where gadolinium oxide (Gd_2O_3) poison, in the form of toroids wrapped around each fuel assembly, is fitted to the first charge of fuel in new reactors. ^{157}Gd has a very large neutron capture cross-section (254 000 barns), but ^{158}Gd — the product of the $^{157}Gd(n, \gamma)$ ^{158}Gd reaction — has a much lower cross-section (2·5 barns); hence gadolinium oxide acts as a burnable poison.

Refuelling in AGRs is generally carried out on-load but with the reactor operating at reduced power, and with an associated economic penalty. The use of Gd_2O_3 in the first charge of fuel enables a larger amount of excess reactivity to be 'built-in' to the reactor when it is first taken to power, by using fuel of higher enrichment. This means that the time period before the reactor first requires refuelling can be increased, thereby reducing the economic penalty associated with refuelling. Similarly, the use of Gd_2O_3 poison in feed fuel (i.e. fuel loaded after the reactor has been producing power for some time), together with increased fuel enrichment, enables the irradiation time or burnup (see Chapter 2) of fuel within the reactor to be increased, and further reduces the economic penalty associated with refuelling.

8.8 Taking a reactor critical

Taking a reactor critical is a routine operation which has to be carried out after each shutdown. Nevertheless the operation has to be performed with some care since, at low powers, the effects of thermal feedback will be at their minimum. Hence the possibility of a power overshoot (as opposed to a gentle power rise) exists. Reactors are designed to protect themselves against this possibility by means of their shutdown systems (see Chapter 11) but nevertheless a 'critical approach' is always undertaken with some care.

The primary objective of a critical approach is thus to take the reactor up to power in a smooth, controlled manner. A secondary objective (particularly following a refuelling shutdown) is to confirm that the 'critical balance height' — the height of the control rods to give k_{eff} exactly equal to unity — is in agreement with that predicted by computer calculations. This then provides a cross-check that any new fuel has been correctly located in the core. In PWRs, for example, it is common to use fuel of different enrichments to help flatten the flux distribution. If a new fuel sub-assembly of 3 per cent enrichment, which would normally be placed at the edge of the core, was wrongly located at the centre of the core where the enrichment is usually about $2 \cdot 5$ per cent, then the critical balance height would be lower than anticipated. This should alert the operators that a mistake had been made. If no action were taken, such fuel would operate at a higher than normal temperature, and perhaps suffer some damage as a result.

If we consider a sub-critical reactor with a neutron source producing S fissions per second and with a multiplication factor k_{eff} ($k_{eff} < 1 \cdot 0$) then we may say that:

S fissions/sec are produced by the source
Sk_{eff} fissions/sec are produced from the previous fission generation
and Sk_{eff}^2 fissions/sec are produced from the generation previous to that.

Hence the total number of fissions per second is:

$$S + Sk_{eff} + Sk_{eff}^2 + \ldots = \frac{S}{1 - k_{eff}}, \text{ for } k_{eff} < 1 \qquad (8.35)$$

and therefore the sub-critical power is ($P_s/(1 - k_{eff})$) watts, where P_s is the power due to source S. Also, we can write that:

$$\rho = \frac{k_{\text{eff}} - 1}{k_{\text{eff}}} \simeq k_{\text{eff}} - 1 \qquad (8.36)$$

if the reactor is just sub-critical. So, for such a reactor:

$$\text{Reactor power} = -\frac{P_s}{\rho} \qquad (8.37)$$

For a typical reactor, the shutdown ('neutron') power P_s is of the order of 50 mW or so.

During a critical approach, the reactivity ρ is increased by raising the control rods. However when $\rho = 0$ (i.e. the reactor is just critical) equation (8.37) becomes meaningless. The control rod height to give a reactivity of exactly zero — the critical balance height — may therefore be determined by extrapolating a line of points plotted on a graph of (1/power) versus (control rod height) as shown in Figure 8.11. Values for reactor power may be obtained from the reactor nucleonic instrumentation (see Chapter 11).

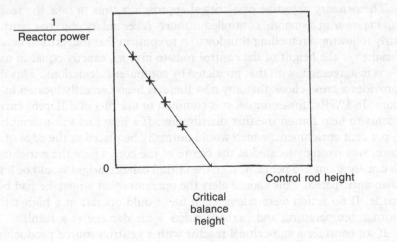

Figure 8.11.   The determination of critical balance height during a critical approach

Chapter 9

Radiation hazards and shielding

One of the most important characteristics of nuclear radiation such as alpha, beta, gamma radiation and neutrons is that these radiations and particles are a hazard to health in human beings and animals due to their ionizing effect in body tissue. Gamma radiation and neutrons, which have a high penetrating power, are an external hazard, and sources of these radiations must be shielded. Alpha and beta particles have a low penetrating power and are only dangerous if a substance emitting these particles is breathed or ingested into the body.

In view of this hazard it is important in the construction of a nuclear reactor to provide adequate shielding so that neutrons and gamma radiation originating in the reactor core are prevented from escaping into the reactor's surroundings where people are working. The same problem arises to a smaller extent with radioactive isotopes which are used on a large scale in science, engineering and medicine. In this case the size and strength of the radioactive source is very much less than the core of a nuclear reactor, but it is nevertheless necessary to reduce the radiation escaping from such a source to an acceptably low level.

In this chapter we will discuss first the effects of radiation on living tissue, the units by which these effects are measured, and the presently accepted limits of radiation dose. The interaction between gamma radiation (the most important radiation from the shielding point of view) and matter will be described in greater detail than was possible in Chapter 1. The shielding of simple radioactive sources will be studied to establish the relationship between source strength, shield thickness and dose rate. The problem of reactor shielding, which involves complex sources of both neutrons and gamma radiation, is beyond the scope of this book and will only be described qualitatively. Finally, the heating of shields due to the absorption of radiation will be discussed.

9.1 Health physics

The study of the effect of nuclear radiation on human beings is known as health physics, and it has become an increasingly important branch of science in recent years as a result of the expanding use of nuclear power as a source of energy, and the widespread use of radioactive isotopes in science, medicine and industry. As a result of these developments there has been a growing public concern about the health effects of the radiation levels which are now present in the environment. Only with a good understanding of the effects of radiation can adequately safe standards of exposure be set, and public fears allayed.

Several different sources of radiation exist in the environment to which the public may be exposed. These include the following:

1. Natural background radiation, which has three components.

   (a) Cosmic radiation, the stream of charged particles (protons and alpha particles) originating in space which bombards the earth's atmosphere and interacts to produce gamma radiation. This gamma radiation reaches the earth's surface, being more intense at high altitude (where the atmosphere is 'thinner') than at low altitude.

   (b) Naturally occurring radioactive isotopes in the earth's crust, such as ^{238}U and ^{232}Th, and their radioactive decay products. The intensity of this source is greater than average in regions of granite rocks, as these contain uranium, and also at one or two places, such as Kerala in India, where the Monazite sands contain quite large quantities of thorium. The most important of the decay products is radon, an inert and heavy gas produced by the decay of both ^{238}U and ^{232}Th. Radon in very small quantities diffuses from uranium and thorium bearing rocks and tends to accumulate in closed spaces, such as unventilated mines, or poorly ventilated buildings. The decay products of radon include radioactive isotopes of lead, bismuth and polonium, and when breathed these can lodge in the lungs and pose a health hazard. Many deaths among early uranium miners were due to this effect.

   (c) Naturally occurring radioactive isotopes, principally potassium 40, in the human body.

2. Medical X-rays and gamma-rays used for diagnostic purposes and the treatment of certain types of disease. Radiation which can cause diseases such as cancer can also be used, in carefully controlled doses, to cure cancer. This is the largest man-made contribution to the radiation dose of the population, but it does not affect the entire population.

3. Miscellaneous natural sources, including the burning of coal which releases very small quantities of uranium and its daughter products into the atmosphere.

4. Fallout from the atmospheric testing of nuclear weapons. This effect is related in terms of time to the atmospheric testing of nuclear weapons.

In the months following tests the radiation level is quite high, and in years when there is no testing the level drops.

5. The discharge of radioactive effluents and the disposal of radioactive waste from nuclear reactors and reprocessing plant. Gaseous fission products, principally krypton, discharged into the atmosphere may be widely disseminated, and liquid waste discharged into the sea may be carried to other places by tides and currents.

6. Occupational exposure, that is the radiation received at work by workers in nuclear power plant, reprocessing plant and other places where radioactive materials are used and handled.

The effect of radiation, such as alpha, beta and gamma radiation and neutrons, is to cause ionization which may damage living cells. As a result, these cells no longer function correctly and radiation sickness or other deleterious effects may result. The extent of the damage depends on the type and intensity of the radiation, the length of time during which it is received, and the total amount of energy of the radiation which is absorbed within the human body.

At this stage the units of radiation and radiation dose will be defined. The unit of radioactivity, the becquerel, has already been introduced and defined as:

$$1 \text{ becquerel (Bq)} = 1 \text{ disintegration/second.}$$

The dose of any radiation is a measure of the energy imparted as a result of ionizing interactions in the material through which the radiation is passing. The original unit introduced for the measurement of doses of X- and gamma radiation is the roentgen (R) which is defined as follows:

The roentgen is that quantity of X- or gamma radiation which produces in one cubic centimetre of air at s.t.p. (0·001 293 gramme) ions carrying a total charge of one electrostatic unit of either sign.

An alternative definition can be deduced by noting that the electronic charge is $4·8 \times 10^{-10}$ electrostatic unit, and that the energy released in the creation of a single ion pair in air is 34 eV. (This figure, although not exact, has been adopted as standard by the International Commission on Radiological Protection.) The energy released in the production of ions with a total charge of one electrostatic unit of either sign is:

$$\frac{34}{4·8 \times 10^{-10}} \text{ eV} = \frac{34 \times 1·6 \times 10^{-19}}{4·8 \times 10^{-10}}$$

$$= 1·133 \times 10^{-8} \text{ joules}$$

This energy is released in $0 \cdot 001\ 293$ gram of air. In one kg of air the energy released due to the same intensity of radiation is:

$$\frac{1 \cdot 133 \times 10^{-8}}{1 \cdot 293 \times 10^{-6}} = 8 \cdot 77 \times 10^{-3} \text{J/kg}$$

The roentgen may thus be defined as the dose of radiation which causes the release of $8 \cdot 77 \times 10^{-3}$ J/kg in air. This definition of the roentgen is based on effects in air whose properties are not the same as those of body tissue. The quantity of gamma radiation which releases $8 \cdot 77 \times 10^{-3}$J/kg in air releases about $9 \cdot 7 \times 10^{-3}$ J/kg in soft body tissue.

The use of this unit of radiation exposure is now more or less discontinued, and it has been described for historical interest.

The present unit of radiation dose which can be applied to any type of radiation is the gray (Gy). It is a measure of the energy of any radiation which is absorbed per unit mass in any material and is defined as:

$$1 \text{ gray} = 1 \text{ joule of absorbed energy/kg}$$

It is known that the energy released (or absorbed) is not the only factor to be considered in estimating radiation damage to the human body. Highly ionizing particles such as alphas or recoil protons (which are produced by the scattering of fast neutrons in hydrogenous materials) are more effective from the point of view of damaging human tissue than beta or gamma radiation, even when the amounts of energy released are in all cases the same. This factor is taken into account by specifying the Quality Factor (QF) of each type of radiation, based on a value of 1 for gamma radiation. The QF for several different types of radiation are given in Table 9.1.

Table 9.1. The Quality Factor of radiations and particles

Radiation	QF
X-, γ-radiation, betas	1
Thermal neutrons	3
Recoil protons, fast neutrons, alpha particles	20
Heavy recoil nuclei	20

The biologically equivalent dose is calculated by multiplying the absorbed dose in grays by the Quality Factor, the unit being called the sievert (Sv). The relationship is:

$$\text{Equivalent dose in sieverts} = \text{Actual dose in grays} \times \text{QF}$$

As an example of this, it is evident that a fast neutron dose of 1 Gy is equivalent biologically to a dose of 20 Gy of gamma radiation.

Returning briefly to the list of radiation sources given above to which the public is exposed, the National Radiological Protection Board (NRPB), the UK agency responsible for matters connected with radiological protection, has estimated that the average annual dose equivalent to members of the UK population in the early nineteen-eighties from all sources was about 2·2 millisieverts (mSv). Of this, natural background radiation contributed 87 per cent, or just less than 2 millisieverts (mSv). Medical radiation contributed 11·5 per cent, and the smallest contribution, about 0·1 per cent, was due to the discharge of radioactive effluents and wastes. (NRPB – R159, 1984). There are, of course, non-typical groups of the population to which the above figures do not apply. For example, the citizens of Aberdeen, the granite city, receive a higher level of natural background radiation due to the traces of uranium in their buildings.

9.2 The biological effects of radiation exposure

The biological effects of exposure to radiation have been studied since the nineteen-twenties when it first became obvious that radiation could cause serious damage to health, and in extreme cases, death. The International Commission on Radiological Protection (ICRP) was established in 1928, and since then it has been the one internationally recognized body responsible for setting standards relating to radiation levels.

In the fifty years of its existence the ICRP has gathered data from many sources, including early workers using radioactive materials in a way which would now not be permitted, survivors of the two atomic bomb explosions over Japan in 1945, people who have received radiation diagnosis and therapy in hospital, workers in the nuclear industry and people living either near to a nuclear plant, such as power stations, or in areas of high natural background radiation. More recently, as a result of the serious reactor accident at Chernobyl in the USSR, another large group of exposed persons, those living near to that reactor at the time of the accident, will provide scientists with much data for years to come on the long-term effects of a reactor accident which resulted in a massive discharge of radioactive material into the environment.

During these fifty years the ICRP has continuously reviewed radiation dose levels that in the light of existing knowledge were regarded as having a negligible probability of causing serious illness. These levels have from time to time been reduced as more evidence has become available, and in 1977 the ICRP recommendation replaced the use of a maximum permissible dose by a more comprehensive system of dose limitation.

The 1977 recommendations may be summarized as:

1. No practice shall be adopted unless its introduction produces a positive net benefit.
2. All radiation exposures shall be kept as low as reasonably achievable (ALARA), economic and social factors being taken into account.

3. The dose equivalent to individuals shall not exceed the limits recommended for the appropriate circumstances by the Commission.

The present whole body dose equivalent limit recommended by the ICRP for members of the public (not including the dose due to natural background radiation and medical sources) is 5 millisieverts per year (mSv/y). In the case of workers undergoing occupational exposure, the corresponding dose limit recommended is 50 millisieverts per year (mSv/y).

The health effects of radiation exposure can be divided into two categories, namely acute and delayed effects. Acute effects are commonly called radiation sickness. The symptoms depend on the magnitude of the dose received. In order of increasing dose the symptoms are fever, internal bleeding, vomiting and diarrhoea, tremors, cramps and coma. The dose threshold for the onset of radiation sickness is about 1 Sv, and for a dose of 4 Sv about 50 per cent of the victims would probably die. Few people could survive an exposure of more than 6 Sv. Radiation sickness may not be evident for up to two weeks following exposure, depending on the dose, and death, if it occurs, may not happen until eight weeks after exposure. In cases of very severe exposure, however, death may follow within two days.

Delayed effects take the form of cancers and hereditary effects, and in discussing them the following terms should be noted:

1. Somatic effects: those effects in which the radiation induced sickness appears in the irradiated person.
2. Hereditary effects: those effects which are evident only in the children of the irradiated person as a result of radiation damage to the reproductive organs.
3. Stochastic effects: those effects for which the probability or occurrence is regarded as a function of the dose of radiation received.

9.3 The effects of inhaled and ingested radioactivity

Studies of the effects of accidents involving the release of radioactivity show that, with the exception of those very close to the scene of the accident at the time it happens, the main hazard to the population as a whole arises from the inhalation or ingestion of air or food which is contaminated by the radioactive material from the accident. For example, the accident at the Russian nuclear power station at Chernobyl in April 1986 produced severe acute somatic effects in a few hundred people, namely the power plant operators and firemen exposed to radiation at the time of the accident and the subsequent fire. In addition, many thousands of people may suffer in years to come from delayed stochastic effects from the radioactive debris of the accident (principally fission products) which was carried in the atmosphere for hundreds of miles from Chernobyl and deposited on the ground by rain, where it entered the food chain of humans and animals. Fallout from nuclear weapons testing produces a similar hazard.

In order to make calculations about the risks of such exposure, it is necessary to consider the following points:

1. How much of each radioactive isotope is inhaled or ingested.
2. The organs of the human body in which each isotope may become concentrated.
3. The dose which each organ receives from a given quantity of inhaled or ingested radioactive material.
4. The dose—risk coefficient, known as the weighting factor, for each organ.

The first point depends on the amount of radioactivity released and its subsequent dispersion in the atmosphere and return to ground level. The second and third points have been the subject of medical research, and the following table shows the organs affected by each of three important radio-active fission products, and the dose (Sv) which results from the inhalation or ingestion of 1 GBq of the fission product concerned.

Table 9.2. The dose due to certain fission products

Isotope	Principal organ(s) affected	Inhalation factor (Sv/GBq)	Ingestion factor (Sv/GBq)
^{90}Sr	(i) Red bone marrow	330	190
	(ii) Bone	730	420
^{131}I	Thyroid	290	480
^{137}Cs	Whole Body	8	13

Source: ICRP 30, Supplement 1, 1979

The weighting factors mentioned above are the ratios of the stochastic risk resulting from radiation damage to an individual organ to the total risk when the body is uniformly irradiated. Values of the weighting factors recommended by the ICRP are given in Table 9.3.

These factors can be used to define the effective dose equivalent as

$$\text{Effective dose equivalent} = \sum_{\substack{\text{all} \\ \text{affected} \\ \text{tissues}}} \left(\begin{array}{l}\text{Dose received by a} \\ \text{particular tissue}\end{array}\right) \times \left(\begin{array}{l}\text{Weighting factor} \\ \text{for that tissue}\end{array}\right)$$

The effective dose equivalent is subject to the same recommendations and limitations as those mentioned earlier for the whole body dose equivalent.

Table 9.3. Weighting factors for individual organs

Tissue	Weighting factor
Red bone marrow	0·12
Bone surface	0·03
Lung	0·12
Thyroid	0·03
Breast	0·15
Gonads	0·25
Remainder	0·30

The most important delayed effect of radiation is cancer, which is serious and may be fatal. Cancer is, however, a relatively common disease from which millions of people die each year. Any person exposed to radiation has an increased probability of contracting cancer, this extra probability being dependent on the radiation dose received. The linear dose−risk relationship assumes (without complete justification) that the increased probability of serious effect such as cancer is proportional to the radiation dose received. This hypothesis enables risk factors to be expressed for exposure of various parts of the body to radiation.

The risk factor expresses the additional number of people in a population sample of stated size who are likely to suffer serious effects if the members of this population are each exposed to an additional dose of 1 Sv. For example, a risk factor of 1 in 200 per Sv or 5×10^{-3} per Sv means that in a group of 200 people each exposed to 1 Sv of radiation to a particular part of their bodies, it is probable that one person will suffer serious effects as a result of this exposure.

The ICRP has made the following recommendations for risk factors for serious hereditary effects and fatal cancers resulting from the irradiation of various tissues and organs.

Table 9.4. Risk factors for irradiation

Tissue or organ	Effect	Risk factor/Sv
Breast	Cancer	$2·50 \times 10^{-3}$
Red bone marrow	Leukaemia	$2·00 \times 10^{-3}$
Lung	Cancer	$2·00 \times 10^{-3}$
Thyroid	Cancer	$5·00 \times 10^{-4}$
Bone surfaces	Cancer	$5·00 \times 10^{-4}$
Other tissues	Cancer	$5·00 \times 10^{-3}$
Whole body, all cancer effects		$1·25 \times 10^{-2}$
Gonads	Hereditary	$4·00 \times 10^{-3}$
Whole body, all health effects		$1·65 \times 10^{-2}$

The data in Tables 9.2, 9.3 and 9.4 can now be used in the following example to calculate the additional probability that a person who drinks milk contaminated with ^{131}I having an activity of 1000 Bq/litre of milk at the rate of $0 \cdot 5$ litres per day for two weeks will develop a fatal cancer as a result.

The dose due to the quantity of contaminated milk is:

$$0 \cdot 5 \times 1000 \times 14 \times 480 \times 10^{-9} = 3 \cdot 36 \times 10^{-3} \text{ Sv}$$

Since ^{131}I is concentrated in the thyroid, for which the weighting factor (Table 9.3) is $0 \cdot 03$, the effective dose equivalent is approximately 10^{-4} Sv.

Using the risk factor (Table 9.4) of $1 \cdot 25 \times 10^{-2}$ for all cancer causing effects, the probability of a person suffering a fatal cancer is

$$10^{-4} \times 1 \cdot 25 \times 10^{-2} = 1 \cdot 25 \times 10^{-6},$$

or approximately one person in a million who drink the contaminated milk is likely to suffer a fatal cancer.

One uncertainty associated with the above calculation is the impossibility of checking its accuracy, since the risk from the dose specified is so small. Any increase in mortality rates from small radiation doses is lost in the normal incidence of death from cancer, which is about 1 in 4 of the population, and only a minute fraction of these deaths is due to natural or man-made radiation.

The allowable quantity of any radioactive isotope which may be taken into the body per year, either by ingestion or inhalation, and results in a whole body effective dose equivalent of $0 \cdot 05$ Sv is called the annual limit of intake (ALI). This quantity is expressed in terms of its radioactivity.

The derived airborne concentration (DAC) is the concentration of a radioactive isotope in air which would result in a person inhaling one ALI in a year. In calculating this quantity it is assumed that the person is occupationally exposed to the radioactive substance in the atmosphere of his working environment which he breathes for 250 days per year at the rate of 10 m^3 per 8 hour working day. Values of annual limits of intake and derived airborne concentrations published by the ICRP are given in Table 9.5.

In concluding this section, it is emphasized that for workers who are occupationally exposed in a radiation environment the total dose rate from external exposure, inhalation and ingestion must not exceed $0 \cdot 05$ Sv/year.

Table 9.5. *Annual limits of intake and derived*
airborne concentrations

Isotope	ALI by ingestion (Bq)	DAC (Bq/m^3)
^{90}Sr	1×10^6	300
^{131}I	1×10^6	700
^{137}Cs	4×10^6	2000

Usually the dose rate is kept well below this level, 0·015 Sv/year being a typical value permitted at nuclear installations. Actual dose rates recorded may be even less. For example, at the fuel processing works of British Nuclear Fuels Ltd at Sellafield and the adjacent Calder Hall nuclear power station the average annual dose of the workers in the years 1977 to 1983 was in the range from 6 to 10 mSv (NRPB − R173, 1984).

Studies of mortality among radiation workers in the UK have shown no detectable excess cancer deaths, in spite of the fact that these workers regularly receive small doses of radiation. Some authors believe that the ICRP assumption of a linear relationship between dose and risk is pessimistic at low doses. It has even been suggested that low doses of radiation might be beneficial to health. A minority of researchers take the opposite view that the ICRP assumption underestimates the risk at low doses.

The effective dose equivalent for members of the public arising from nuclear installations is kept below 1 mSv per year per person for the most exposed group of individuals, and the dose received by the public as a whole from this source is much less than this level, being about 0·002 mSv/year per person in the early nineteen-eighties. This figure may be compared with natural background radiation which, as stated earlier, leads to an average dose of 2 mSv/year per person.

9.4   The interaction of gamma radiation and matter

The three processes known as Compton scattering, pair production and the photoelectric effect have already been mentioned in Chapter 1. The rates at which these processes occur depend on the cross-sections or absorption coefficients for each process. These cross-sections are similar to neutron cross-sections and will be considered in detail shortly.

At low gamma ray energies (less than 0·5 MeV) the most important process is the photoelectric effect. In this interaction the photon interacts with an orbital electron and transfers all its energy to it. The electron is ejected from its orbit and the photon disappears and may be considered to be absorbed. The photoelectric cross-section decreases rapidly as the photon energy E_γ increases, being roughly proportional to $E_\gamma^{-7/2}$. The cross-section also depends on the atomic number of the target material, being proportional to Z^n, where the index n has (depending on the photon energy and the element in question) a value between 3 and 5. In general the photoelectric effect is most important with heavy elements and low photon energies.

The pair production process occurs with gamma radiation of energy greater than 1·02 MeV, and the photon disappears with the production of a positron and an electron. The positron is shortlived, and when it annihilates itself with an electron two gamma photons of 0·51 MeV energy are formed. These can generally be neglected from the shielding point of view because of their low energy and isotropic emission. Thus pair production can be regarded as an absorption process like the photoelectric effect. The threshold energy for pair production is 1·02 MeV, which is equivalent to the mass of a positron and an electron. The cross-section is roughly proportional to Z^2. In view of this, it is evident that pair production is more important in heavy than light elements.

In the Compton scattering process it may be assumed that a photon collides with an orbital electron and that as a result of the collision another photon of lower energy emerges. The difference between the energies of the incident and emergent photons is transferred to the electron. This is not an absorption reaction as a photon of reduced energy remains after the interaction. The Compton scattering cross-section per atom is inversely proportional to the energy E_γ in the range 0·5 to 2 MeV, and is proportional to the atomic number Z of the material.

The total cross-section per atom for gamma ray interaction is the sum of the three individual cross-sections:

$$\sigma_t = \sigma_{PE} + \sigma_{PP} + \sigma_{CS} \qquad (9.1)$$

The energy absorption cross-section, which gives a measure of the rate at which the energy of gamma radiation is attenuated, is the sum of the photoelectric and pair production cross-sections plus a fraction of the Compton scattering cross-section which represents the average fraction f of the photon energy lost in this process:

$$\sigma_{tE} = \sigma_{PE} + \sigma_{PP} + f\sigma_{CS} \qquad (9.2)$$

The fraction f may be determined by analysing the Compton scattering process.

The cross-sections per atom are analogous to microscopic neutron cross-sections. It is more usual to use absorption coefficients which are analogous to macroscopic neutron cross-sections. Thus, if the number of atoms per cubic centimetre of an element is N, these absorption coefficients are:

the linear absorption coefficient, $\mu = N\sigma_t$; and
the linear energy absorption coefficient, $\mu_E = N\sigma_{tE}$.

The units of linear absorption coefficients are cm^2/cm^3 or cm^{-1}.

The cross-sections per unit mass are known as mass absorption coefficients, and are obtained by dividing the linear absorption coefficients by the material density (g/cm^3). Thus the mass absorption coefficient is $\chi = \mu/\rho$, and the mass energy absorption coefficient is $\chi_E = \mu_E/\rho \ \mathrm{cm^2/g}$.

The rate of interaction of gamma radiation in matter is given by equations that are similar to those for neutron interactions. Consider a collimated beam of gamma radiation of intensity ϕ_γ photons/cm² s incident on a target material of thickness dx and area A normal to the beam, containing N atoms/cm³. The following equations apply:

the rate of interaction in the target $= N\sigma_t \phi_\gamma A \ \mathrm{d}x$,
the rate of interaction per unit volume $= N\sigma_t \phi_\gamma = \mu\phi_\gamma$,
the rate of energy absorption per unit volume $= \mu_E \phi_\gamma E_\gamma$,

where E_γ is the energy of the gamma radiation:

the rate of interaction per unit mass $= \chi\phi_\gamma$; and
the rate of energy absorption per unit mass $= \chi_E \phi_\gamma E_\gamma$.

The attenuation of the beam of gamma radiation is given by the equation:

$$\phi_\gamma(x) = \phi_{\gamma 0} \, e^{-\mu x} \tag{9.3}$$

where $\phi_\gamma(x)$ is the flux of radiation of initial intensity $\phi_{\gamma 0}$ which penetrates a distance x into the target material without interacting. This equation assumes that all interaction processes result in the gamma radiation being removed from the original beam.

Figure 9.1 shows the variation of the individual and total absorption coefficients for aluminium and lead, which are typical light and heavy elements. It is of interest to note that for radiation of about 1 or 2 MeV

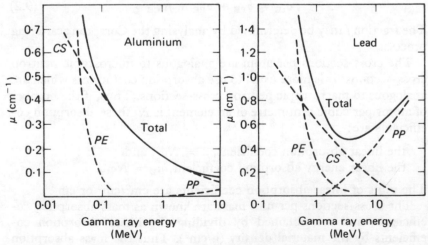

Figure 9.1. Gamma radiation absorption coefficients for aluminium and lead

energy, the Compton scattering process predominates for both elements, and the total cross-section is approximately equal to the Compton scattering cross-section. The Compton scattering cross-section is proportional to the atomic number Z, which for most elements is approximately half the atomic mass (hydrogen and the very heavy elements are exceptions). According to Avogadro's Hypothesis the number of atoms per unit mass of any element is inversely proportional to its atomic mass. Consequently the Compton scattering cross-sections per unit mass, and thus the mass absorption coefficients, of most elements are nearly equal for gamma radiation of about 1 MeV energy. The value is about $0.06 \, \text{cm}^2/\text{g}$. The same is true for mass energy absorption coefficients, whose value at 1 MeV is about 0.027 cm^2/g. From this it is evident that equal masses of all materials have approximately the same shielding effect for gamma radiation of 1 or 2 MeV energy.

9.5 The shielding of sources of gamma radiation

Consider a point source of gamma radiation, for example a small capsule of a radioactive isotope, emitting S photons per second isotropically. The absorption of radiation in air may be neglected because of the very low density of air, and if the source is unshielded, the flux of gamma radiation at a distance r from the source is:

$$\frac{\text{The number of photons emitted per second by the source}}{\text{The surface area of a sphere of radius } r}$$

Therefore
$$\phi_\gamma = \frac{S}{4\pi r^2} \tag{9.4}$$

If the source is shielded by a spherical shield of thickness x (see Figure 9.2), then according to equation (9.3) the fraction of gamma photons which pass through the shield without interacting is $e^{-\mu x}$. Combining

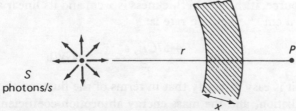

S
photons/s

Figure 9.2. Shielding of a point source of radiation

equations (9.3) and (9.4), the flux of gamma radiation at a point P distant r from a source emitting S photons per second, shielded by a thickness x of material whose linear absorption coefficient is μ is given by the equation:

$$\phi_\gamma = \frac{S}{4\pi r^2}\, e^{-\mu x} \qquad (9.5)$$

This equation neglects the effect of multiple scattering in the shield as a result of which a photon after one or more scattering collisions may contribute to the flux at P. This effect will be discussed later, however equation (9.5) is valid if the thickness of the shield is small enough for multiple scattering to be insignificant.

The dose rate due to a point source of gamma radiation may now be determined. Consider an unshielded source of strength C curies emitting one photon per disintegration of energy E_γ. The flux at a distance 1 metre from the source is, using equation (9.4):

$$\phi_\gamma = \frac{3\cdot7 \times 10^{10} \times C}{4\pi \times 10^4}$$

$$= 2\cdot94 \times 10^5 \times C \quad \text{photons/cm}^2\,\text{s}$$

Taking a value of $0\cdot027$ cm^2/g for χ_E, which is approximately correct for most materials with gamma radiation of energy about 1 MeV, the rate of energy absorption per gramme is:

$$0\cdot027 \times 2\cdot94 \times 10^5\, CE_\gamma$$

$$= 7\cdot94 \times 10^3\, CE_\gamma \text{ MeV/g s}$$

Expressed in the S.I. unit of absorbed dose, this expression becomes:

Absorbed dose rate $= 7\cdot94 \times 10^3 \times 3600 \times 1\cdot6 \times 10^{-13}\, CE_\gamma$
$\qquad\qquad\qquad = 4\cdot57 \times 10^{-3}\, CE_\gamma \text{ Gy/h}$
$\qquad\qquad\qquad = 4\cdot57\, CE_\gamma \text{ mGy/h}$

This equation gives the dose rate 1 metre from an unshielded point, source of radiation of strength C curies and energy E_γ MeV per photon. A more general expression for the dose rate due to a shielded point source is obtained from equation (9.5). At a distance r metres from the source, if the shield thickness is x cm and its linear absorption coefficient μ cm^{-1}, the dose rate is:

$$D = \frac{4\cdot57 CE_\gamma\, e^{-\mu x}}{r^2} \quad \text{mGy/h}$$

Finally, it is easy to verify that in terms of the flux and the energy of gamma radiation, and the mass energy absorption coefficient, the dose rate is given by the equation:

$$D = 5\cdot76 \times 10^{-4}\, \phi_\gamma E_\gamma \chi_E \quad \text{mGy/h}$$

Equation (9.5) for a point source of radiation can be used to solve problems involving more complicated source geometries. Let us con-

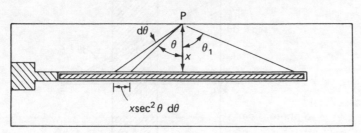

Figure 9.3. A shielded fuel element — a line source of radiation

sider, for example, an irradiated fuel element which after undergoing fission in a reactor for a long time is intensely radioactive due to the build up of fission products in the fuel. Such a fuel element might be a rod 2·5 cm diameter and 1 metre long, and it can be regarded as a finite line source of radiation. Figure 9.3 shows the fuel element enclosed in a container of wall thickness x.

If the activity of the source is S photons/cm s, and, referring to Figure 9.3, the element of length of the source is $x \sec^2 \theta \, d\theta$, then the flux of gamma radiation at the joint P on the surface of the container is given by:

$$\phi_\gamma = 2 \int_0^{\theta_1} \frac{S \times \sec^2 \theta \, d\theta}{4\pi x^2 \sec^2 \theta} \, e^{-\mu x \sec \theta}$$

$$= \frac{S}{2\pi x} \int_0^{\theta_1} e^{-\mu x \sec \theta} \, d\theta \tag{9.6}$$

$\int_0^{\theta_1} e^{-\mu x \sec \theta} \, d\theta$ is known as the Secant Integral, sec $i \, (\mu x, \theta_1)$ and values of this integral for various values of θ_1 and μx are tabulated in the radiation shielding literature.

Finally, we will consider an infinite plane source of radiation emitting S photons/cm^2 s isotropically, with an infinite slab shield of thickness x, see Figure 9.4.

The area of a 'ring element' of the source subtended by an angle $d\theta$ at P is $2\pi r^2 \sin \theta \sec \theta \, d\theta$, and the flux of gamma radiation at point P due to the 'ring source' is given by:

$$d\phi_\gamma = \frac{2\pi S r^2 \sin \theta \sec \theta \, d\theta}{4\pi r^2} \, e^{-\mu x \sec \theta}$$

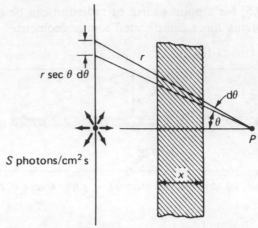

Figure 9.4.   An infinite plane source of radiation

The flux of gamma radiation at P due to the entire plane source is given by:

$$\phi_\gamma = \frac{S}{2} \int_0^{\pi/2} \tan\theta \; e^{-\mu x \sec\theta} \; d\theta$$

If $y = \mu x \sec\theta$, $dy = \mu x \tan\theta \sec\theta \; d\theta$, and the preceding equation for the flux becomes:

$$\phi_\gamma = \frac{S}{2} \int_{\mu x}^{\infty} \frac{e^{-y}}{y} \; dy \tag{9.7}$$

$\int_{\mu x}^{\infty} (e^{-y}/y) \; dy$ is known as the Exponential Integral, $E_1(\mu x)$, and values of this integral for various values of μx are tabulated in shielding literature.

The preceding equations (9.5), (9.6) and (9.7) refer to the uncollided flux of gamma radiation at the point of measurement P. If the beam is broad, or the shield thick enough to cause multiple scattering, the actual flux is greater than the uncollided flux due to the fact that Compton scattering may result in some photons contributing to the flux at P after one or more such collisions. This effect is known as build-up, and allowance for it is made by the introduction of build-up factors.

The flux build-up factor, B_F, is defined by the equation

$$B_F = \frac{\text{Total gamma flux at the point of interest}}{\text{Uncollided gamma flux at the point of interest}}$$

and the equation for the attenuation of a parallel beam of gamma radiation, (9.5), is modified to become:

$$\phi_\gamma = \phi_{\gamma_0} e^{-\mu x} B_F$$

The dose build-up factor, B_D, is defined by the equation

$$B_D = \frac{\text{Total dose at the point of interest}}{\text{Dose due to uncollided gamma radiation only}}$$

The determination of build-up factors depends on the following points:

(i) the type of build-up of interest (flux or dose),
(ii) the material through which the radiation is passing,
(iii) the thickness of the absorber,
(iv) the energy of the gamma radiation,
(v) the source geometry (point or distributed).

It is possible by the numerical analysis of multiple scattering collisions to calculate build-up factors, and the results of these calculations are shown in Figure 9.5 for water and lead. The abscissa of these

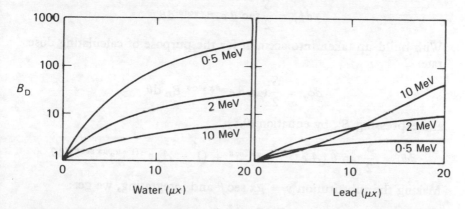

Figure 9.5. Build-up factors for a point source of radiation in water and lead

graphs are in units of (μx), the number of mean free paths through the absorber. It is noticeable that the lighter material, water, has in general larger build-up factors because Compton scattering (which is the process responsible for build-up effects) is more significant in light than heavy elements.

From the results of the numerical analysis it has been found possible to express build-up factors as the sum of two exponential terms by the equation:

$$B_D = A e^{-\alpha_1 \mu x} + (1 - A) e^{-\alpha_2 \mu x} \qquad (9.8)$$

where A, α_1 and α_2 are constants for any material and gamma ray energy, and are selected to fit the numerical results referred to above. Equation (9.8) may be incorporated into the uncollided flux equations;

for example, for the purposes of calculating dose rate, equation (9.5) becomes:

$$\phi_\gamma = B_D \frac{S}{4\pi r^2} e^{-\mu x}$$

$$= \frac{S}{4\pi r^2} \{A e^{-(1+\alpha_1)\mu x} + (1 - A) e^{-(1+\alpha_2)\mu x}\} \qquad (9.9)$$

The equation for the flux with build-up is thus similar to that for the uncollided flux, with amended exponents and weighting factors.

This formulation is very convenient if integration is carried out over a distributed source, for example the infinite plane source considered earlier. For such a source the uncollided flux at point P due to an elementary 'ring source' was previously shown to be given by (refer to Figure 9.4).

$$d\phi_\gamma = \frac{S}{2} \tan \theta \, e^{-\mu x \sec \theta} \, d\theta$$

With build-up taken into account for the purpose of calculating dose rate:

$$d\phi_\gamma = \frac{S}{2} \tan \theta \, e^{-\mu x \sec \theta} \, B_D \, d\theta$$

and expressing B_D by equation (9.8):

$$d\phi_\gamma = \frac{S}{2} \tan \theta \{A e^{-(1+\alpha_1)\mu x \sec \theta} + (1 - A) e^{-(1+\alpha_2)\mu x \sec \theta}\} \, d\theta$$

Making the substitution $y = \mu x \sec \theta$ and integrating, we get:

$$\phi_\gamma = \frac{S}{2} \int_{\mu x}^{\infty} \sum_{i=1}^{2} \frac{A_i}{y} e^{-(1+\alpha_i)y} \, dy$$

$$= \frac{S}{2} \sum_{i=1}^{2} A_i E_1(\mu_i x) \qquad (9.10)$$

where $E_1(\mu_i x)$ is the Exponential Integral previously introduced, and $A_1 = A$, $A_2 = (1 - A)$, $\mu_1 = (1 + \alpha_1)\mu$, and $\mu_2 = (1 + \alpha_2)\mu$.

Comparing equations (9.9) and (9.10) with (9.5) and (9.7), it is evident that the general rule for the application of build-up factors of the form given by equation (9.8) is:

If the uncollided flux is expressed by the equation:

$$\phi_\gamma = F(r)G(\mu x)$$

where $F(r)$ is a function of the distance from the source to the point of measurement, and $G(\mu x)$ is a function of the shield thickness, then the total flux at the same point is given by:

$$\phi_\gamma = F(r) \sum_{i=1}^{2} A_i G\{(1 + \alpha_i)\mu x\} \qquad (9.11)$$

As an example of the application of build-up factors, we will consider an irradiated fuel element from a nuclear reactor, stored under water in a cooling pond. The radioactive fission products emit gamma radiation with a wide spectrum of energies, however we will simplify matters by considering radiation of a single energy.

Example: An irradiated fuel element 70 cm long, which may be regarded as a uniform line source of radiation, is stored under 3 m of water. The activity of the fuel element is 2000 curies, and the energy of the emitted radiation is 2 MeV per photon. Calculate the dose rate at the surface of the water immediately above the fuel element.

For gamma radiation of 2 MeV energy, μ for water $= 0.0493$ cm^{-1}, and the values of the constants in equation (9.8) are $A = 6.4$, $\alpha_1 = -0.076$ and $\alpha_2 = 0.092$.

For the given geometry (see Figure 9.3):

$$\tan \theta_1 = \frac{35}{300} = 0.117 \qquad \theta_1 = 6.7°$$

and
$$\mu x = 300 \times 0.0493 = 14.8$$

$$\phi_\gamma = \frac{2000 \times 3.7 \times 10^{10}}{70 \times 2\pi \times 300} \times$$

$$[6.4 \sec i\{(1 - 0.076)14.8, 6.7°\} - 5.4 \sec i\{(1 + 0.092)14.8, 6.7°\}]$$

The values of the Secant Integrals are tabulated in handbooks of reactor shielding:

$$\sec i(13.68, 6.7°) = 1.3 \times 10^{-7}$$

and
$$\sec i(16.18, 6.7°) = 1.1 \times 10^{-8}$$

Using these values:

$$\phi_\gamma = 434 \text{ photons/cm}^2 \text{ s at the surface of the water}$$

The mass energy absorption coefficient for 2 MeV radiation in air is 0.023 cm^2/g, and the dose rate is (using the equation derived earlier):

$$D = 5.76 \times 10^{-4} \times 434 \times 2 \times 0.023$$

$$= 0.0115 \text{ mGy/h}$$

9.6 Reactor shielding

The problem of reactor shielding is a great deal more complicated than that of shielding a simple source of gamma radiation. In a reactor the principal source of radiation, neutrons and gamma radiation, is in the core itself, however radioactive coolant flowing in ducts and heat exchangers, and irradiated fuel elements in storage pits may provide additional sources of gamma radiation.

Neutrons and gamma radiation produced in the core have a wide spectrum of energies, and neutrons passing from the core into the shield produce more gamma radiation as a result of (n, γ) reactions in the shield. Beta radiation is produced by fission product decay, but only high energy beta radiation is important because as it is slowed down, gamma radiation known as Bremsstrahlung (braking radiation), is produced.

The important sources of radiation in a reactor can be summarized as follows:

1. *Neutrons.* The characteristics of prompt fission neutrons and delayed neutrons have been described in Chapters 2 and 8.

2. *Prompt gamma radiation.* This is the gamma radiation emitted at the instant of fission; the average number of photons emitted per fission is 8 with an energy spectrum up to about 7 MeV. The average energy per photon is about 1 MeV.

3. *Fission product gamma radiation.* This is the gamma radiation emitted during the decay of radioactive fission products. The radiation from short-lived fission products is important during reactor operation and rapidly decays when the reactor is shut down. The radiation from long-lived fission products may be of more importance after shutdown, particularly in the cooling of fuel in the core and the shielding of irradiated fuel elements.

4. *Capture gamma radiation.* This radiation is emitted as a result of (n, γ) reactions in the core and (to a lesser extent) in the shield. It provides a source of radiation in the shield itself.

5. *Activation product gamma radiation.* If the product of an (n, γ) reaction is radioactive, it will provide a further source of gamma radiation as it decays. Reactor coolants which become radioactive as a result of (n, γ) reactions in the core are a source of radiation in external ducts and heat exchangers both during operation and after shutdown. Sodium is an important example of such a coolant.

Less important sources of radiation are:

1. *Gamma radiation*—from the inelastic scattering of fast neutrons.

2. *Photo-neutrons*—resulting from (γ, n) reactions in reactors containing significant quantities of heavy water of beryllium. (^{2}H and ^{9}Be are the isotopes responsible for this reaction.)

3. *Annihilation radiation*. This is the name given to the 0·51 MeV gamma radiation which results from the annihilation of positrons with electrons. The positrons may originate from high energy gamma radiation undergoing pair production processes.

4. *Bremsstrahlung*. This gamma radiation is produced by the deceleration of high energy beta particles. It may be of importance if ^7Li is used as a reactor coolant because neutron capture in ^7Li produces ^8Li which decays by the emission of high energy beta particles.

The main radiation shield of a reactor is known as the biological shield. Its function is to reduce the intensity of neutrons and gamma radiation escaping from the reactor to a level which is acceptable from the health physics point of view. The design of the biological shield involves the following steps:

1. Determine the neutron flux distribution in the core of the reactor, and hence determine the distribution of sources of gamma radiation in the core, and the energy spectrum of these sources.
2. Determine the current and energy of neutrons leaking out of the core into the shield, and hence the rate of slowing down and absorption of neutrons in the shield. This establishes the sources of gamma radiation in the shield.
3. The neutron current through the outer surface of the shield is determined.
4. Once the sources of gamma radiation in the core and shield are known, the gamma flux and energy in the shield, and the current through its outer surface are determined.
5. The dose rate at the outer surface of the shield is calculated from the neutron and gamma currents through it. The shield thickness is designed so that this dose rate is within acceptable limits.
6. The rate of energy absorption of gamma radiation and neutrons in the shielding must be determined. This energy absorption results in the heating of the shield, and in the case of power reactors this heating may lead to unacceptable temperatures and thermal stresses.

To reduce the heating effect in the biological shield, a comparatively thin thermal shield is placed between it and the core of the reactor. The function of the thermal shield, which is cooled by a suitable flow of coolant, is to reduce the neutron and gamma currents from the core into the biological shield, and thereby reduce the rate of energy absorption and the temperature rise in this shield. The thermal shield should have a good thermal conductivity, and be able to withstand thermal stresses. Steel is commonly used as the material of the thermal shield.

The biological shield should contain some hydrogen compound to slow down fast neutrons, and be dense enough to attenuate gamma radiation effectively. Concrete satisfies both these requirements fairly well and is suitable for land-based reactors. Barytes concrete, containing the heavy element barium, and steel-shot concrete have been used for biological shields. They are more dense than ordinary concrete, with improved shielding properties, however their higher cost offsets this advantage. The biological shield for a marine reactor, which is usually a fairly compact pressurized water reactor, must satisfy a minimum space and weight requirement. This leads to a shield design which consists typically of alternate layers of water (for fast neutron slowing) and steel (for gamma ray attenuation).

9.7 Shield heating

As we have seen, the slowing down of neutrons and the absorption of gamma radiation are energy absorbing processes in which the energy lost by the radiation is transferred to the material through which it is passing, and results in an increase in the temperature of the material. The first step in determining the temperature rise in a reactor shield involves a knowledge of the distribution of the energy sources, i.e. the spatial variation and the energy spectrum of neutrons and gamma radiation in the shield. Once the distribution of energy sources is known, the temperature variation in the shield may be found by application of the general heat conduction equation, equation (6.5), with suitable boundary conditions.

As a simple illustrative example we will consider the effect of a parallel beam of gamma radiation passing through a slab-shaped shield. The spatial variation of the energy source, assuming exponential attenuation of the radiation, is given by:

$$H(x) = 1 \cdot 6 \times 10^{-13} \, \mu_E E_\gamma \phi_{\gamma 0} \, e^{-\mu x} \ \text{W/cm}^3$$

the x coordinate being normal to the face of the shield.

The heat conduction equation for this problem is (neglecting temperature variation in the y and z directions):

$$\frac{\mathrm{d}^2 T}{\mathrm{d} x^2} = - \frac{C}{k_s} e^{-\mu x} \tag{9.12}$$

where $C = 1 \cdot 6 \times 10^{-7} \, \mu_E E_\gamma \phi_{\gamma 0}$, and k_s is the thermal conductivity of the shield material, W/m °C.

Equation (9.12) is integrated twice, and the boundary conditions applied that $T = T_w$, the temperature of the inner and outer surfaces of the shield when $x = 0$ and L, the thickness of the shield. The solution for T is:

$$T - T_w = \frac{C}{\mu^2} \left[(1 - e^{-\mu x}) - \frac{x}{L} (1 - e^{-\mu L}) \right] \tag{9.13}$$

The magnitude of this effect can be illustrated by considering a concrete shield 2 metres thick with an incident energy flux of gamma radiation of 10 mW/cm². (These figures are typical of a power reactor of the Calder Hall type which has a 15 cm thick steel thermal shield inside the concrete biological shield.) The linear absorption coefficient and thermal conductivity of concrete are taken as 0·085 cm⁻¹ and 2·0 W/m °C respectively, and it is assumed that all gamma interactions are absorption processes. If equation (9.13) is evaluated for these figures, the result for the maximum concrete temperature is:

$$T_{max} - T_w = 4·3°C$$

This temperature difference is quite acceptable from the point of view of thermal stresses. However, to consider again the example of the Calder Hall reactor, if there were no thermal shield then the energy flux entering the biological shield would be about five or ten times greater, and the maximum temperature difference would be about ten times greater than the figure just calculated, namely about 40°C. This temperature difference would cause a significant thermal stress in the concrete shield, and it is to prevent this that the thermal shield is used.

Chapter 10

Materials for nuclear reactors

The development of nuclear power in the last fifty years has involved the use of many materials — metals, alloys, ceramics, liquids and gases — which in pre-nuclear power years were relatively unknown to engineers and little used in engineering. Now these materials are becoming well known, and although not exactly commonplace, they are part of the vast range of materials used in engineering today.

It is the purpose of this chapter to give brief and qualitative description of the materials which have acquired special importance in nuclear engineering, describing their properties and the role that they play in nuclear reactor technology.

The physical characteristics of materials are of considerable importance in determining their use in nuclear reactors. Characteristics such as strength, hardness, ductility, melting point, boiling point, density and thermal conductivity are all familiar to engineers in their choice of materials for particular applications. Not so well known (before the advent of nuclear power) was the neutron cross-section, which is of crucial importance in the selection of most materials for use in the core of nuclear reactors. Most components of a reactor are subjected to high temperatures and stresses, and their properties under such conditions must be taken into consideration.

It is also necessary to have an understanding of any possible chemical reactions that may occur between materials in a reactor. For example, the coolant comes into contact at high temperature with several other components of the reactor and heat exchangers, and any possible reactions such as corrosion involving these components must be well understood.

One type of change affecting materials in reactors which is not familiar in non-nuclear applications is radiation damage. This is the name for changes which can be caused to the molecular and crystalline structure of a material as a result of its being exposed in a neutron flux, particularly a fast neutron flux, for prolonged periods. Equally, structural damage can be caused within the fuel of a reactor as a result of bombardment of individual atoms by high energy fission products. Radiation damage causes the displacement of individual atoms within the crystals of a material and changes the crystalline structure; this effect may be anisotropic, leading to expansion along some axes and contraction along others. The results of this damage may include cracking, swelling and creep as well as changes in other properties such as ductility, hardness, strength and thermal conductivity. These results are usually only of significance after long periods of irradiation, and much research and testing in materials testing reactors, extending over several years, is needed to determine the nature and extent of radiation induced damage.

It is possible to list a number of criteria by which reactor materials may be compared and selected. These criteria include the following:

1. Good mechanical properties including (where necessary) high thermal conductivity, specific heat, density, strength, ductility, melting or boiling point and low coefficient of expansion.
2. Low neutron capture cross-section is necessary for all materials in the core except the fuel and the control rods (and burnable poisons if used).
3. Chemical stability of all materials at the operating temperatures and pressures of the reactor. No risk of oxidation, decomposition, explosion or other chemical reaction.
4. No metallurgical phase changes at operating temperatures that may lead to dimensional changes.
5. Resistance to significant radiation damage within the lifetime of the material in the reactor.
6. Materials chosen should be readily obtainable in a pure form, cheap, easy to fabricate and non-toxic.

10.1 Fuels

Uranium

Uranium in one form or another is by far the most common fuel material for nuclear reactors. (By comparison, the use of thorium and plutonium has so far been on a very small scale.) It can be used either as pure uranium, a metal, or as a compound such as uranium dioxide UO_2 or uranium carbide UC.

Uranium is a rather soft and ductile metal which oxidizes readily in air and water at high temperature. Its melting point is 1133°C. It exists in one of three allotropic forms, depending on its temperature. These three different forms are called the alpha, beta and gamma phases, and changes from one phase to another due to temperature changes are accompanied by density changes. Alpha phase uranium has a density of 19 g/cm^3 and a thermal conductivity which varies from 25 W/mK at 25°C to 42 W/mK at 665°C. The transition from the alpha to the beta phase takes place at 665°C and is accompanied by dimensional changes in the crystalline structure of the uranium, expansion along one axis and contraction along the others. To avoid distortion due to these anisotropic dimensional changes 665°C is considered to be the maximum operating temperature for uranium.

Metallic uranium is also very susceptible to radiation damage which produces dimensional changes and swelling above about 450°C. Consequently high burnups of metallic uranium fuel are not possible. In the British gas-cooled Magnox reactors, which are the principal users of this type of fuel, the burnup is limited to about 3500 MWd/t.

To summarize, the low operating temperature, susceptibility to radiation damage and low permissible burnup of uranium are serious disadvantages to its choice as a reactor fuel, and account for its very limited use.

Uranium dioxide UO_2 is a black powder which can be fabricated by cold pressing and sintering at high temperature to produce small cylindrical pellets, and in this form it is by far the most common material for the fuel of commercial reactors. In this ceramic form UO_2 has good stability at high temperature and good resistance to radiation damage which enables it to be used to high burnups. The melting point is 2865°C and the theoretical density is $10 \cdot 96$ g/cm^3, although in practice the density of UO_2 pellets produced as described above is about 10 g/cm^3. The thermal conductivity is low, being about $2 \cdot 5$ W/mK in the temperature range from 1000 to 2000°C, however this low thermal conductivity is compensated for by the very high melting point which permits high maximum fuel temperatures.

Uranium dioxide does not react with water at high temperatures, a very valuable characteristic as otherwise cladding failures in water cooled reactors would lead to serious reactions. It can retain a large fraction of the gaseous fission products at temperatures below 1000°C, but as the fuel temperature at the centre of a pellet is likely to be greatly in excess of this value, provision must be made for fission product gas release. This is usually done by having an empty space at the top of each fuel tube into which the gases can diffuse.

During operation in a reactor UO_2 pellets suffer structural changes, principally as a result of the high operating temperatures and high temperature gradients, but also as a result of prolonged irradiation. The effects may include swelling, formation of cracks and voids in the pellet and changes in the grain structure of the UO_2. This type of fuel is normally subjected to much higher burnups than pure uranium, and 5 per cent or more of the original uranium atoms in the fuel may undergo fission and be changed, each one to two intermediate mass fission product atoms.

Uranium carbide, UC, is another ceramic fuel of possible interest, but it has not been developed or used to anything like the same extent as UO_2. It may have some advantages over UO_2, principally its higher thermal conductivity and higher density which leads to more uranium atoms per unit volume of fuel, which is an advantage in a reactor. Uranium carbide reacts with water, which makes it unsuitable for use in water cooled reactors, but it does not react with sodium below 500°C, so it might be used in fast reactors. Its melting point, 2380°C, is rather lower than that of UO_2, but this is compensated for by its higher thermal conductivity. The development of uranium carbide so far has been principally as the fuel for high temperature gas-cooled reactors.

Plutonium

Pure plutonium metal is not suitable as a reactor fuel due to the large number of crystalline phases which exist up to its melting point of 640°C. The thermal conductivity is also very low, about 4·2 W/mK at room temperature. Plutonium metal is highly reactive in moist air, but it can be stored in dry air at low temperature. It is a very dangerous material, being radioactive, toxic and an essential component of nuclear weapons, and is potentially a serious health hazard, particularly if it exists as dust in the atmosphere and is taken into the lungs by inhalation.

As a reactor fuel plutonium is used as the oxide PuO_2. Its melting point is 2400°C. Plutonium dioxide is mixed with uranium dioxide to form mixed oxide fuel (MOX) which for fast reactors typically contains 20 to 25 per cent of PuO_2. The properties of this mixed oxide fuel are similar to those of UO_2 alone.

Thorium

Thorium has not been used as a reactor fuel to any great extent yet except in a few high temperature gas-cooled reactors. Thorium 232 is the fertile isotope from which uranium 233 is produced as described in Chapter 3, and it is theoretically possible to obtain high breeding ratios in thermal as well as fast reactors using this combination.

Pure metallic thorium has a melting point of about 1700°C. It is superior to uranium due to its better stability, but it is not used as a fuel in its pure form. Instead it is used either as thorium dioxide ThO_2 or thorium carbide ThC_2. To date these compounds have only been used to a very small extent in a few high temperature gas-cooled reactors.

Thorium dioxide is similar in many respects to uranium dioxide. It is produced by the same methods of powder metallurgy, and it is chemically inert and has a good resistance to radiation damage. Thorium carbide has been used in the form of coated particle fuel in HTGRs. Very small spherical particles less than 1 mm diameter of mixed ThC_2 and UC_2 (highly enriched in ^{235}U) are coated with thin layers of pyrolitic carbon and silicon carbide to retain fission products. These particles are dispersed in graphite to form a homogeneous mixture of fuel and moderator which has a very high operating temperature and good resistance to radiation damage.

10.2 Moderators

The requirements of the moderator for a thermal reactor, namely low mass number, very low neutron capture cross-section and high scattering cross-section, limit the choice to only a few materials. Hydrogen and its isotope deuterium, carbon and beryllium are the only elements that are suitable. Hydrogen and deuterium, being gases, are not sufficiently dense and must be used in the form of compounds, water and heavy water being the obvious choices. The use of hydrocarbon compounds has been tried, but has not been successful and such materials are not used as moderators. It is interesting to recall, however, that Fermi used paraffin wax in his early experiments in the 1930s to slow down neutrons and study their interactions with the elements, so he was one of the first scientists to be aware of the effects of neutron moderation.

Beryllium has a very low neutron capture cross-section (0·009 barns), high melting point (about 1300°C) and good strength, and at one time it seemed possible that it would find an application either as the moderator or the fuel cladding in thermal reactors. However, it and its compounds are toxic, and beryllium itself has low ductility and poor corrosion resistance. Beryllium oxide BeO also has undesirable properties. As a result of this neither beryllium nor its oxide have found any use in power reactors, and it is unlikely that they will be used in the future.

The choice of moderators for thermal reactors is thus limited to three materials — water, heavy water and carbon in the form of graphite.

Water

Water is an obvious choice for the moderator of a thermal reactor, and it can also serve as the coolant. It has excellent neutron slowing down properties which enable water moderated reactors to have much more compact cores than are possible in other types of thermal reactors. The capture cross-section of water is rather high (0·66 barns per molecule) so that water moderated and cooled reactors require enriched uranium for criticality. It is, of course, abundant, cheap and easily obtainable with high purity.

The main problem associated with the use of water as the moderator and coolant in a power reactor concerns its rather unfavourable thermodynamic characteristics. The saturation pressure and temperature relationship is such that high pressures are required to prevent boiling at high temperatures, e.g. a pressure of 150 bar is required to allow water to reach a temperature of 340°C without boiling. Pressures of 150 to 160 bar are typical of pressurized water reactors, in which the water temperature is limited to about 325°C.

It is important to maintain water purity in a water cooled and moderated reactor, firstly to minimize corrosion and secondly to prevent the water from becoming radioactive due to (n, γ) reactions with the impurities as the water flows through the reactor core. Radiation levels in the water can influence the radiation dose levels to which power station operating and maintenance staff are exposed, and the maintenance of high water purity assists in reducing operator exposures.

Heavy water

Heavy water is similar to ordinary water in most of its physical and thermo-dynamic properties. The principal difference is that deuterium has a much lower capture cross-section than hydrogen, and the capture cross-section of heavy water is only $0 \cdot 001$ barns. Deuterium is, however, not such a good moderator as hydrogen. Consequently, heavy water moderated and cooled reactors use natural uranium as their fuel, but their core sizes are rather larger than water moderated reactor cores. An important difference is that heavy water is very expensive to produce by separation from ordinary water, and losses by leakage must be minimized.

Graphite

The world's first nuclear reactor, CP-1, (Chicago Pile 1) was moderated with graphite, and although this material has not been used subsequently in American commercial power reactors, it has been used extensively in British reactors. Its nuclear characteristics, i.e. slowing down power and capture cross-section are not as good as those of heavy water, but it is readily obtainable in a pure form at reasonable cost and is easily machined. Its structural and thermal properties are good although it reacts with oxygen and air at high temperatures. Graphite has a high thermal conductivity (130– 190 W/mK) and it sublimes at 3650°C without melting, so there is for all practical purposes no limit on its maximum operating temperature. Graphite cores are assembled from large numbers (many thousands) of blocks of rectangular shape in which holes are provided for fuel elements and control rods.

The effect of prolonged neutron irradiation on graphite is very important, as such irradiation causes dimensional changes and the buildup of stored energy within the crystalline structure. The dimensional changes are due to the anisotropic crystalline structure of graphite, and they depend on its temperature during irradiation. Below 300°C there is a contraction along one axis and expansion along another. At higher temperatures there is contraction in all directions. The design of a graphite core and the keys which hold one block in position relative to its neighbours must make these dimensional changes possible without damage or distortion of the core.

The stored energy, sometimes called Wigner energy, is due to the energy required to displace atoms in the graphite's crystalline structure. Neutron irradiation causes the displacement of atoms, and if they do not return to their original positions, this energy remains stored in the graphite. This is the case at temperatures below 100°C. At higher temperatures partial self-annealing of the graphite takes place and some of the stored energy is released. It is necessary to release the stored energy in a controlled way otherwise, if it is released too rapidly, the graphite temperature can rise disastrously. This occurred at one of Britain's two original air-cooled plutonium producing reactors at Windscale, Cumbria in 1957. On that occasion an uncontrolled Wigner energy release caused the graphite temperature to reach such a high level that the graphite burned and the core of the reactor was destroyed.

Present British gas-cooled reactors operate with graphite temperatures of about 400°C at which there is continuous self-annealing of the Wigner strain energy, and no possibility exists of a repeat of the Windscale accident.

10.3 Coolants

The principal requirements of the coolant for a nuclear reactor are as follows:

1. Good thermodynamic properties, namely high thermal conductivity, density and specific heat, and low viscosity.
2. Chemically non-reactive with other components of the reactor.
3. Very low neutron capture cross-section.
4. It should not become radioactive as a result of (n, γ) reactions which may occur when the coolant is passing through the core of the reactor.

Among the gaseous coolants, some can be eliminated from consideration for one reason or another. Oxygen and hydrogen are both reactive, the latter explosively so. Nitrogen has a significant capture cross-section ($1 \cdot 8$ barns). Air, being a mixture of oxygen and nitrogen can also be ruled out. Oxygen 16 undergoes an (n, p) reaction with high energy neutrons (e.g. fission neutrons) to form nitrogen 16 which is radioactive, but its half-life is only 7 seconds, so the radioactive hazard is short-lived. The two most suitable gaseous coolants are carbon dioxide and helium.

Carbon dioxide is a fairly unreactive gas, but it does react at high temperatures with certain types of steel and with graphite. Both these reactions have proved troublesome in British gas-cooled reactors in which carbon dioxide has been chosen as the coolant. Its advantages are its inertness, availability and cheapness, and the very low capture cross-sections of both carbon and oxygen. Carbon 13, which is a very small constituent of naturally occurring carbon, captures neutrons to a very small extent to form radioactive carbon 14 which poses a minor hazard if carbon dioxide leaks or is vented from a reactor to the atmosphere.

Helium is inert, has good thermodynamic properties and does not pose a radioactive hazard, so it might be regarded as the ideal gaseous reactor coolant. Unfortunately, it is not readily available in large quantities and is expensive. Its use as a reactor coolant is at present confined to the few high temperature gas-cooled reactors operating in the USA and West Germany.

The properties of water and heavy water as reactor coolants have been described in the preceding section, but their moderating ability makes them unsuitable for use as coolants in fast reactors.

Liquid metals are potentially excellent reactor coolants because of their good thermodynamic properties, in particular their high thermal conductivity which leads to very good heat transfer coefficients. Sodium, lithium, mercury and sodium–potassium alloys are all possibilities, but of these only sodium has been used to any great extent, exclusively in fast breeder reactors. Sodium–potassium alloys may become more commonly used. Mercury is very expensive and is toxic, as well as having too high a capture cross-section for use in thermal reactors. Lithium is similar to sodium in many respects, but has a higher melting point and is more expensive.

Sodium is the standard coolant for the still fairly small number of fast breeder reactors operating in the world. Its melting point is 98°C and its boiling point at atmospheric pressure is 890°C, so it is not necessary to use sodium at higher than atmospheric pressure, which is a distinct advantage. It is highly reactive with air and water, so high integrity pipework and heat exchangers are necessary to avoid leakage. Sodium has a fairly low capture cross-section (0·5 barns) but it does undergo the $^{23}Na(n, \gamma)\ ^{24}Na$ reaction, and intermediate heat exchangers are required to contain the radioactive sodium 24 within the biological shield of the reactor, as described in Chapter 7 (section 7.10). Sodium is not corrosive to most structural materials provided its oxygen content is maintained low. The formation of sodium oxide in the coolant circuit can lead to plugging unless it is removed in cold traps.

Sodium–potassium eutectic alloy with a melting point of −11°C can be used for the decay heat removal systems of fast reactors as the coolant temperature in these systems may fall below the melting point of sodium.

10.4 Cladding materials

The fuel cladding materials require a very low neutron capture cross-section, good thermal conductivity and good strength at high temperatures to resist thermal stresses, deformation of the fuel and build up of fission product gas pressure inside the cladding. In addition the cladding should be easy to fabricate and not liable to corrosion or other chemical reactions with either the fuel or the coolant. The following metals and their alloys satisfy some if not all the above requirements, and have been considered for use as fuel cladding: aluminium, beryllium, magnesium, zirconium and stainless steel.

As pointed out in section 10.2, beryllium has not proved to be a suitable material for use in reactors, and it has not been developed as fuel cladding.

Aluminium has been used as the cladding material for low power research reactors in which its temperature seldom exceeds about 100°C. Its capture cross-section is fairly low (0·23 barns) but its mechanical properties such as strength and hardness are quite low and aluminium cannot be used in situations of high stress or temperatures greater than 300°C. Consequently it cannot be used as a cladding or structural material in power reactors, and its use as noted above is solely in research reactors, in which it is often used in the form of an inter-metallic compound with uranium.

Magnesium has a very low capture cross-section (0·063 barns) but like aluminium it is a soft metal of low strength with a maximum operating temperature of about 450°C. It is also very reactive, oxidizing readily in air. The characteristics of magnesium particularly as regards oxidation, can be greatly improved by the addition of small quantities (less than 1 per cent) of aluminium and beryllium, and the alloy thus formed, called Magnox, has been used extensively as the cladding material in British gas-cooled reactors. However, the maximum operating temperature of Magnox is about 450°C, and this places a limit on the thermodynamic performance of these reactors.

Zirconium has a low capture cross-section (0·185 barns), a high melting point (1850°C), good mechanical properties and a high resistance to corrosion by water. These properties make it an excellent cladding material for reactors. The mechanical properties and corrosion resistance can be further improved by alloying zirconium with small quantities of tin (1·5 per cent), iron (0·15−0·2 per cent), chromium (0·1 per cent) and nickel (up to 0·05 per cent). The resulting alloys, Zircaloy-2 and Zircaloy-4, are used extensively as the cladding and in-core structural materials for water and heavy water moderated and cooled reactors. They are the pre-eminent fuel cladding materials at the present time.

At very high temperatures, above about 1000°C, zirconium and the Zircaloys can react with steam to form hydrogen, which is potentially very hazardous. The conditions which might give rise to this reaction can occur after a reactor loss-of-coolant accident when the fuel and cladding overheat as a result of inadequate decay heat removal. If the cladding temperature exceeds 1000°C and the cladding comes into contact with steam, then the hydrogen producing reaction occurs, possibly with explosive results.

Stainless steel is well known for its excellent mechanical properties and corrosion resistance. Unfortunately its capture cross-section, which depends on the type of steel and the precise quantities of chromium and nickel in it, is too high for stainless steel to be regarded as ideal for fuel cladding or other in-core uses. It is not now used as fuel cladding in pressurized water reactors, but it is used in the advanced gas-cooled reactor and in fast breeder reactors. In the latter the rather high capture cross-section is of less importance than in thermal reactors. Stainless steel is used extensively for out-of-core uses in PWRs and FBRs where its excellent corrosion resistance is invaluable.

10.5 Control materials

Materials for controlling reactors need to have high capture cross-sections. Several such materials are available. In addition other materials of lower cross-section can be used in the reactor core for 'flux-shaping' and fine control. Stainless steel rods can be used for this purpose.

Of the control materials, boron (σ_c = 760 barns) is the most common. It cannot be used by itself, but may be alloyed with steel or used in the form of boron carbide encased in steel. Boron can also be used to compensate for long term changes of reactivity (such as fuel burnup) in the form of a burnable poison, boric acid, dissolved in the coolant of a pressurized water reactor.

Indium and cadmium both have high capture cross-sections (195 and 2450 barns respectively) but both have melting points which are too low to permit their use in power reactors. In the form of an alloy containing 80 per cent silver, 15 per cent indium and 5 per cent cadmium, these elements are used as control rods in pressurized water reactors. The alloy has an adequately high melting point, but must be encased in stainless steel to protect it from corrosion.

Hafnium (σ_c = 113 barns) is a good control material, having adequate mechanical strength and good corrosion resistance. However, it is too expensive to be used on a large scale for commercial reactors.

Gadolinium (σ_c = 46 000 barns) is used as a burnable poison in some reactors such as the advanced gas-cooled reactor.

Chapter 11

Safety and hazards in nuclear power

From the earliest years of nuclear reactor development it has been realized that an accident in which a reactor becomes supercritical by a substantial amount (i.e. greater than the delayed neutron fraction), resulting in a sudden rise in the fission rate, power and temperatures, is an event that can have very serious consequences. These consequences might include fuel overheating and meltdown, release of radioactive fission products into the coolant and possibly eventually into the atmosphere, and very rapid increases in temperatures and pressures resulting in possible rupture of pressure vessels and other containment structures.

As a result reactor designers have paid great attention to the inherent safety of reactors which can be achieved by negative temperature and power coefficients and fail-safe control systems. It can be said with some confidence that present-day thermal reactors are safe in the sense that under no conceivable circumstance can they explode like a bomb, and control systems have been designed which can, in the event of any malfunction on the part of the reactor or its associated plant, automatically and rapidly shut down the reactor, i.e. make it subcritical by a substantial amount, in a very few seconds.

Fast reactors, with their greater concentration of fissile fuel, would appear to be inherently more hazardous than thermal reactors, and it is conceivable that if the core of a fast reactor melted the fuel might slump into a supercritical configuration which would result in a very sudden energy release, but hardly an explosion. Needless to say, fast reactors are designed in such a way as to eliminate this possibility, and a negative temperature coefficient is an important characteristic for a fast reactor.

Many reactor-years of operating experience have shown that it is not the fission chain reaction in the reactor core that is the most likely source of malfunction and accidents, but the 'conventional' components of the power plant such as pumps, valves, switches, relays and parts under stress such as pressure vessels or pipework. Human error on the part of operating and maintenance staff has also proved to be a rather frequent source of trouble in nuclear power plant.

These factors are not peculiar to nuclear power plant, but they assume great importance because of the hazardous nature of nuclear reactors. Designers have to ensure that all systems should as far as possible be fail-safe and redundant, i.e. if one system fails to function correctly, another is available to fulfil the same function.

As stated above, nuclear reactors cannot explode like nuclear bombs. This is primarily because of the fast acting negative thermal feedback due to Doppler broadening of the ^{238}U absorption resonances, as described in Chapter 8. In addition in thermal reactors where neutrons are moderated, the prompt neutron lifetime l_p is of the order of 10^{-4} seconds; in a bomb, since the neutrons are unmoderated, the prompt neutron lifetime is of the order of 10^{-8} seconds. Finally, reactor fuel consists typically of 2 to 3 per cent ^{235}U, whereas nuclear weapons contain almost pure ^{239}Pu. The net effect of these differences is that, even in a reactor which is totally out of control and has gone prompt critical (as happened at Chernobyl), the reactor period will not be much less than a second or so. In a nuclear bomb, the period is of the order of nanoseconds. (The reactor period has been defined in Chapter 8.)

The hazard from nuclear power lies instead in the possibility of an accident that might release large quantities of radioactivity to the atmosphere. To prevent such an occurrence, nuclear reactors must satisfy each of three criteria:

1. There must be reliable and rapid shutdown of the reactor if required under all foreseeable circumstances.
2. The decay heat of the fuel, i.e. the heat which continues to be produced by the reactor fuel after shutdown due to fission product decay, must be reliably rejected to the environment.
3. There must be reliable, sound containment of the fission products.

Each of these criteria will be examined further below.

11.1 Reactor shutdown systems

The primary means of shutting down most designs of reactor is the interruption of the electrical supplies to the electromagnetic clutches which suspend the control rods above the core. The control rods then fall into the core under gravity (Figure 11.1). (An exception to this is the BWR design, where the control rods enter the core from below. Here the rods are driven into the core hydraulically in the event of a shutdown.)

It is of great importance for the safe operation of nuclear power stations that reliable, automatic shutdown should occur whenever certain parameters exceed their design margins. This important task is performed by the Automatic Protective System (APS). The APS on a typical reactor might monitor the following parameters:

Turbine trip
Boiler feedwater flowrate
Neutron flux ϕ
Core outlet temperature
Reactor period or the rate of increase of the neutron flux $d\phi/dt$
Cladding failure (leading to increased levels of radioactivity in the coolant)
Coolant flow rate
Neutron flux/coolant flow ratio
Coolant level or pressure

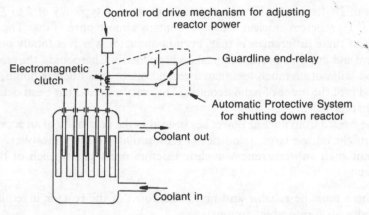

Figure 11.1.   Schematic illustrations of reactor Automatic Protective System

It is normal practice for each parameter to be monitored by three instruments. Each instrument is connected to a trip amplifier, which goes to a 'tripped' state if the signal exceeds design margins. If two of the three trip amplifiers go to the tripped state, this de-energizes the 'guardline end-relay' (Figure 11.2) and interrupts the current to the electromagnetic clutches.

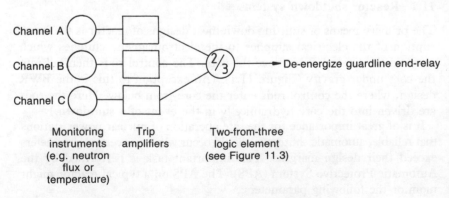

Figure 11.2.   Majority voting shutdown system

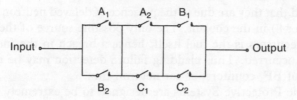

Figure 11.3. Relay logic for a two-from-three majority voting system. Relays A_1 and A_2 receive signals from the trip amplifiers of channel A, etc (see Figure 11.2). A trip signal from any two channels will interrupt the connection between input and output

By use of such 'majority voting' systems, the high reliability of reactor shutdown is assured while the probability of spurious shutdown is minimized.

The transducers for measuring, say, temperatures and flowrates are conventional thermocouples and flowmeters. However, the methods for measuring neutron flux and detecting cladding failure are unique to the nuclear industry and will now be described briefly.

For the measurement of the neutron flux various types of instrument exist, of which one type is the boron trifluoride or BF_3 counter. This instrument is similar to the conventional ionization chamber, but is filled with BF_3 gas in which neutrons cause the $^{10}B(n, \alpha)\ ^7Li$ reaction, which in turn produces alpha particles. These particles may then be detected and counted as in a conventional ionization chamber. Because of the wide power range that has to be monitored (typically from 30 mW at shutdown to 3000 MW at full power, a range of 10^{11}) it is normal practice to use three different types of neutron flux measuring instruments, each one covering a part of the whole power range. Thus there will normally be three low power instruments, three medium power instruments and three high power instruments, each operating on majority voting logic. As the power changes during startup or shutdown, the operators must switch the protective system from one power range to the next; failure to do so will cause a reactor trip.

The detection of fuel cladding failure is important, especially in gas-cooled and fast reactors, since the cladding forms the first barrier against the release of fission products. Furthermore, cladding failure may be an indication that the fuel is suffering abnormal conditions; for example a partial blockage of coolant flow in a fuel assembly leading to localized higher-than-normal temperatures. The detection of cladding failure is also important from the viewpoint of maintenance since, if an excessive amount of fission products are released into the primary coolant circuit, the resulting radioactive contamination may lead to unacceptable dose rates to the maintenance staff at a subsequent shutdown.

Cladding failure is detected by monitoring samples of coolant for neutrons away from the core and behind neutron shielding. If neutrons are detected, it is inferred that they are due to the presence of delayed neutron precursors (see Chapter 8) in the coolant. The only possible source of these delayed neutron precursors is the fuel itself; hence a breach in the fuel cladding must have occurred. Thus cladding failure detection may be carried out by means of BF_3 counters, as for neutron flux.

Automatic Protective Systems are designed to be extremely reliable — typically better than 10^{-6} failures per annum, which is the same as saying one failure per million reactor operating years. As well as having majority voting, or 'redundant', logic as described above, the APS design normally incorporates 'diversity' also. This means that the APS consists of two separate trains of transducers, trip amplifiers and logic elements, each using different operating principles and each capable, independently of the other, of de-energizing the control rod clutches. Diversity in design ensures that the possibility of 'common-mode' failures — where similar components might fail simultaneously due to, say, some foreseeable external factor such as high temperature, or else some unforeseen disturbance — is minimized. Thus the diverse protective system might incorporate, say, different means of detecting loss of coolant flow (such as pump speed instead of pump pressure rise), a different make of trip amplifier, and an alternative type of logic element such as an electromagnetic switching device called a 'laddic'. (Microprocessor logic elements are not yet in common use, largely due to uncertainties in quantifying their reliabilities. This situation is likely to change as experience in their use is accumulated.)

Finally, most reactors have a diverse, or secondary, means of shutting down. This may take the form of injecting borated water into the coolant (in the case of water-cooled reactors) or nitrogen injection (in the case of gas-cooled reactors).

11.2 Fission product decay heating

A very important characteristic of a nuclear reactor, particularly one that has been operating for a long time, is that although the control system can shut down the chain reaction very rapidly, with the delayed neutrons causing a few minutes' delay before fission finally dies out, there is still a considerable energy release from the fuel after shutdown due to the radioactive decay of the accumulated fission products. This energy must be transferred from the core by a continuing circulation of coolant, otherwise the fuel temperature will rise, resulting in meltdown or high temperature reactions such as the zirconium-steam reaction which occurs at temperatures greater than 1200°C. This particular reaction produces hydrogen which adds yet another hazard.

A precise and detailed description of the decay of the many fission products (whose half-lives vary from fractions of a second to many years) is possible, but the resulting expression is too complex to be of practical use. A simplified expression for the energy release due to the decay of the radioactive fission products can be stated as follows:–

The rate of energy release dP_{fp} by fission product decay at any instant due to a reactor having operated at a power P for an interval of time dt, t seconds earlier is given by:

$$dP_{fp} = 0 \cdot 012 P t^{-1 \cdot 2} dt \tag{11.1}$$

This equation is not precise, its accuracy is to within 50 per cent. Using it and postulating a reactor operating at a steady power P for a time t_o, the rate of energy release due to fission product decay at a time t_s after shutdown, $P_{fp}(t_o, t_s)$, can be found by integrating equation (11.1) from t_s to $(t_s + t_o)$ as Figure 11.4 makes clear.

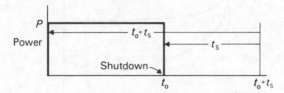

Figure 11.4.   Illustration of the equation for fission product heating

$$\text{Thus: } P_{fp}(t_o, t_s) = 0 \cdot 012 P \int_{t_s}^{t_o + t_s} t^{-1 \cdot 2} \, dt$$

$$= 0 \cdot 06 P \left[t_s^{-0 \cdot 2} - (t_o + t_s)^{-0 \cdot 2} \right] \tag{11.2}$$

This equation is valid for values of t_s greater than about 10 seconds. At this time after reactor shutdown the fission product power is about 4 per cent of the reactor power before shutdown. In a high power reactor this may be quite a considerable level, and it will decay only as the fission products decay. Nothing can be done to control this rate of energy release, and as already mentioned it is essential to remove the heat generated both while the fuel is in the reactor core and after it has been removed for storage in cooling tanks.

To illustrate the effect of fission product heating, Figure 11.5 shows the variations of the ratio P_{fp}/P after the shutdown of a reactor which has operated (i) for a very long time $(t_o \rightarrow \infty)$, and (ii) for 50 days.

A study of Figure 11.5 shows that the fission product decay power ratio is nearly the same in both cases soon after shutdown, because at this time it is the short half-life fission products $(T_{\frac{1}{2}} < 1$ day) which contribute most to the decay power, and these fission products have reached an equilibrium concentration in the reactor prior to shutdown regardless of whether it has been operating for 50 days or much longer. Several days after shutdown the decay power is due to the long half-life fission products which in a reactor which has only operated for 50 days have not had time to reach equilibrium concentration. The activity in such a reactor is much less than in a reactor which has operated for a very long time, in which the long

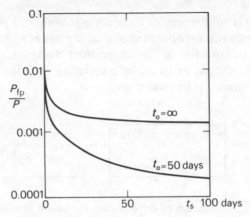

Figure 11.5. Fission product decay heating after shutdown

half-life fission products have reached equilibrium concentration before shutdown.

It follows that an important safety feature of any power reactor is that after shutdown it must be possible to continue to circulate the coolant through the core to remove the decay heat. In the event of failure of power supplies to coolant circulating pumps (which would automatically shut the reactor down immediately) there should be emergency power supplies available; if pumps fail altogether, natural circulation of coolant through the core should be sufficient to remove the decay heat, and this is a feature of gas-cooled reactors.

In PWRs, decay heat rejection is assured by means of circulation of the coolant, with heat rejection via the boilers to the turbine condenser; also, a diverse system of decay heat rejection draws water from the primary circuit and passes it through separate, dedicated decay heat rejection heat exchangers. Finally, and in the event of a loss-of-coolant, there is an emergency core cooling system (ECCS) which injects borated water directly into the core.

In gas-cooled reactors, decay heat rejection is achieved by the circulation of the primary coolant to dedicated auxiliary boilers, and hence to the turbine condenser; alternatively, the steam may be rejected straight to the atmosphere.

In fast reactors, decay heat can be rejected to the atmosphere via natural circulation thermosiphons filled with a sodium—potassium alloy; this is only necessary if heat rejection via the boilers and the turbine condenser is not possible.

11.3 Containment

The containment of fission products and fuel is the final requirement for safe reactor operation. The barriers to the release of fission products are as follows:

1. The fuel element cladding: This forms the first barrier to the release of fission products particularly in gas-cooled and fast reactors. The failure of the cladding in a fuel element may be detected by the means discussed in section 11.1. Thus the primary coolant circuit of these reactors should normally be free from fission products, although there will be other radio-active species present due to neutron activation. Operation with some failed claddings may sometimes be permitted in water-cooled reactors.
2. The reactor pressure vessel and the primary coolant circuit.
3. For water-cooled reactors, a containment building must be erected around the reactor primary circuit (Figure 11.6). This is primarily because, otherwise, it could be envisaged that a failure of the primary circuit might lead to cladding failures (through loss of coolant pressure and fuel overheating), and hence a release of fission products to the environment. (In gas-cooled reactors, most postulated failures of the pressure vessel and primary coolant circuit would not lead to fuel cladding

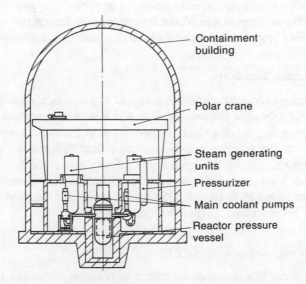

Figure 11.6. A typical PWR containment building layout

failures. Hence there is no release of fission products, and no requirement for a containment building. In addition, all modern gas-cooled reactors use reinforced concrete pressure vessels; sudden catastrophic failure of such vessels is extremely unlikely.)

The containment building should be designed to withstand possible explosive disruption, which could occur if other very serious faults have already arisen:

1. In water-cooled reactors, the possibility of 'steam explosions' should be considered. This is the name given to the event when extremely hot material, at a temperature of several thousand degrees, comes into contact with water. The water may flash into steam with explosive violence. Such an event could occur if fuel somehow reached temperatures very much greater than normal, as happened at Chernobyl. (Steam explosions are sometimes known as 'foundry explosions', since such accidents occasionally occur in iron foundries.)
2. Also in water-cooled reactors, the zirconium fuel cladding will react with steam at temperatures greater than about 1200°C, producing hydrogen. A confined explosion within the containment building might then be envisaged which could conceivably rupture the building if it was sufficiently strong. Such an event would require the fuel to be uncooled so that it could reach the temperatures required. This happened at Three Mile Island (see next section), but the containment building withstood the overpressure caused by the hydrogen combustion.
3. Sudden catastrophic failure of steel pressure vessels has very occasionally occurred in non-nuclear systems. The probability of this happening in reactor systems is minimized by careful non-destructive examination and proof testing of the pressure circuit, and strict quality control.

11.4 Reactor accidents

Serious reactor accidents, which might lead to a significant release of radioactivity, can only arise following events which may be described under four broad categories, as described below. In a well-designed system, each accident route requires a number of coincident failures for a release of radioactivity to occur; the probability of any given accident is thereby minimized. Two serious reactor accidents which have occurred in the last decade, at Three Mile Island, USA in 1979 and at Chernobyl, USSR in 1986 will be described in this section.

Reactivity insertion accident (RIA)

This can occur if a significant amount of reactivity is added to a reactor core. Such an event could arise due to an uncontrolled withdrawal of control rods from the reactor due to faults in their control systems, with a coincident APS failure. This event could also arise in a reactor design which has fast-acting positive reactivity coefficients (leading to instability), with a coincident APS failure and/or a poor fast shutdown capability. (The last scenario occurred at Chernobyl.) The probability of such accidents is minimized by ensuring that all fast-acting reactivity coefficients are negative, and that reliable, fast-acting, diverse shutdown systems exist. These accidents cannot lead to nuclear explosions.

Loss-of-coolant accident (LOCA)

This arises as a result of a breach of the primary coolant circuit, with a coincident failure of the emergency core cooling system (ECCS). In particular, the term LOCA usually relates to water-cooled reactors, where a complete loss of coolant will also give rise to a loss of decay heat rejection capability, and hence fuel damage. The Three Mile Island accident was such an event. Gas-cooled reactors are not subject to LOCAs, but to the less severe event of a depressurization of the primary circuit; decay heat rejection would normally be sustained in such accidents.

The term *anticipated transient without scram* (ATWS) is used in water-cooled reactor systems to refer to the event which would follow a 'routine' plant failure – say a turbine trip or a feedwater pump trip – when the reactor trip systems fail to insert the control rods into the core. Such events should not lead to any activity release in any well-designed systems; the negative reactivity coefficients should lead to a rapid fall in reactor power, and, in any case, most power reactors are fitted with secondary shutdown systems (see section 11.1).

Certain *external events* could conceivably lead to reactor accidents, say an earthquake or collision from a crashing aircraft. Reactors are commonly designed to resist the maximum magnitude of earthquake that is foreseeable in the district where the plant is to be sited. Similarly, reactors are often designed to resist light aircraft crashes and they are usually sited away from civil airline or airforce training routes. Hence the risk from external events can be minimized.

The accident at Three Mile Island

The Three Mile Island Power Station is situated on an island in the Susquehanna River ten miles from the town of Harrisburg in Pennsylvania. The power station had two pressurized water reactors whose features are similar to those described in Chapter 7. The layout of the reactor and its associated plant are shown in Figure 11.7.

The accident, which affected only one of the two reactors, originated with a failure of the feedwater system which caused the feedwater pump A to

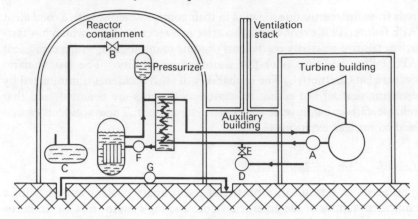

Figure 11.7. Layout of the Three Mile Island PWR

cut out. The turbine and the reactor shut down automatically, but with the feedwater pump out of action, fission product decay heat was not transferred from the reactor via the heat exchangers. As pressure in the reactor and the primary system increased, the pressure relief valve B opened and primary coolant was released to the pressurizer relief tank C. At the same time the standby feedwater pumps D started up, but were unable to deliver water to the heat exchangers because valves E were closed, contrary to operating regulations. About 8 minutes after the start of the accident the operators opened valves E and flow was established in the secondary circuit; at the same time the pressure in the primary system continued to drop because the safety valve B was stuck in the open position. As the reactor pressure dropped, the ECCS was automatically started (as it is designed to do) and cold water was injected into the reactor. A few minutes later an instrument indication of rising water level in the pressurizer led the operators to shut off the ECCS.

An hour or two later severe vibration in the primary circuit pumps F led the operators to switch off these pumps and circulation in the primary systems stopped. Meanwhile with decay heat still being produced, the water in the core boiled and part of the core was uncovered for a few hours, with the steam environment unable to provide adequate cooling. The resulting high temperature caused damage to the fuel elements resulting in release of certain volatile fission products into the primary system, and these fission products found their way via the pressurizer, the still open pressure relief valve, the pressurizer relief tank (which overflowed) into the sump in the reactor containment floor. From there the contaminated water was transferred by the pump G into the auxiliary building and there the radioactive volatile fission products vented from the water and escaped into the atmosphere via the ventilation stack. Thus significant quantities of xenon 135 and iodine 131, with smaller amounts of krypton, escaped into the atmosphere; the iodine release, which was the most significant in radiological terms, amounted to some 16 curies.

Meanwhile, back in the reactor the very high fuel temperatures in the uncovered part of the core caused the zirconium-steam reaction and hydrogen was formed, collecting in the top of the pressure vessel and causing a restriction to the flow of primary coolant. It was thought at the time that this hydrogen presented an explosion risk, but this is unlikely as oxygen was not present in the pressure vessel. Hydrogen did, however, escape into the containment building, and ten hours after the beginning of the accident there was a hydrogen 'burn' which caused a rise in the pressure within the containment building.

About 16 hours after the start of the accident the primary coolant pumps were restarted and water flow through the reactor was re-established to cool the core. Gradually during the next few days the hydrogen in the pressure vessel was removed and the accident was brought under control. There can be no doubt that, although there were no immediate disastrous consequences, this was a very serious accident resulting in the release of considerable quantities of radioactivity into the atmosphere and the destruction of the reactor itself, which is now out of action.

The lessons to be learned from this accident are many. It certainly illustrates how a relatively simple fault can lead through a sequence of events to a major accident, and how this sequence of events can occur as a result of a combination of malfunction of everyday engineering components such as pumps and valves, human error and misjudgement and inadequate design.

It has been estimated that this accident may lead to one additional cancer death among the nearby residents in the years to come. The financial cost of one destroyed reactor and the subsequent clean-up operation has been huge.

The accident at Chernobyl

In April 1986, the No. 4 reactor of the Chernobyl power station, situated on Pripyat river north of Kiev in the USSR, suffered a major accident which caused a significant release of radioactivity to the environment. The reactor was of the RBMK graphite-moderated, water-cooled, pressure-tube design, which is a design unique to the USSR. The layout of the plant is shown in Figure 11.8.

The reactor has many features which are not found in other reactor systems, and which raised concern about its safety. A team of British engineers visited the Soviet Union in the 1970s and reported the following concerns about the RBMK design (amongst others):

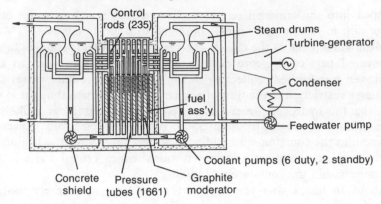

Figure 11.8.   Layout of the RBMK reactor at Chernobyl

1. The reactivity worth of the control rods was inadequate to meet all possible shutdown requirements.
2. The reactor had a positive void coefficient and a positive power coefficient at low powers. This arose because the design was over-moderated which, in turn, was apparently because a decision had been made to use fuel that was only about two per cent enriched, since the USSR was short of enrichment capacity.
3. There was no secondary shutdown system.
4. The graphite moderator normally operated at temperatures in excess of 700°C, at which it spontaneously ignites on exposure to air.
5. The containment structure was less robust than would normally be the case for Western water-cooled reactors.

Furthermore, we now know that the design allowed greater freedom of action for the operators than might be considered normal in a Western design. For example, the operators were apparently able to override trip systems at the flick of a switch; in Western designs, key interlock systems prevent such precipitate action. Also it was essential for the safe operation of the plant that the control rods should never be withdrawn beyond the point at which the control rod 'reactivity margin' became dangerously low; this vital aspect was left entirely to the operators, with no automatic trip system.

Finally, it would appear that reactor emergency shutdowns were normally carried out by motoring the rods into the core, and not by disconnecting electromagnetic clutches. Because the rods were driven-in instead of being allowed to drop, shutdown could take 20 seconds to occur.

Ironically, the accident occurred during a test to check whether the emergency core cooling systems were adequate. The test was to be performed during a routine shutdown. The main events were as follows:

1. Power was to be reduced in stages from 3200 MW(th) to 700 MW(th). Operator error caused the power to undershoot to 30 MW, and the consequent increase in xenon poisoning meant that the operators could only manage to raise the power back to 200 MW. This power level was lower than stipulated in the test instructions — the operators should have abandoned the test at this point.
2. In violation of the test procedure, the operators started up the standby coolant pumps. This meant that core power was 7 per cent normal and coolant flow was 120 per cent normal. This made the whole core virtually isothermal.
3. Difficulties in steam drum water level control at this juncture led the operators to override the reactor trip signals generated by low drum level. They then topped up the drum water level under manual control, with relatively cold feedwater. This caused a fall in reactor temperature, and hence (via the positive void coefficient) the control rods had to be withdrawn yet further to maintain reactor power level at 200 MW(th).
4. The operators noticed that the control rod reactivity margin was too low. Indeed, the reactivity margin was at a level where operating rules stipulated that the reactor should have been shut down. No such action was taken.
5. The test was to be initiated by tripping the turbine. Normally, this would have tripped the reactor also. However, it would appear that the operators wanted to reserve their options in case the test of the emergency core cooling system was unsatisfactory; hence they overrode the reactor trip. This meant (they thought) that they would be able to repeat the test if necessary.
6. At 01.23.10 h on the 26th April 1986, the turbine was tripped. The operators had contrived, unwittingly, to put an already unstable reactor design into a highly dangerous condition. The reactor was operating at low power, and was almost isothermal due to the high coolant flow-rate. By tripping the turbine the only significant heat sink had been removed. In addition, the control rods were too far out of the core to have any significant immediate effect in the event of a reactor trip.
7. The coolant temperature now began to rise steadily until it approached the saturation temperature, at which point bulk boiling occurs. Because of the virtually isothermal state of the reactor, the value of the positive void coefficient had been maximized, i.e. bulk boiling would begin throughout the core more-or-less simultaneously.
8. At 01.23.40 h, a rise in power was noted and a reactor trip was initiated manually, by starting to drive the control rods into the core. Because the rods were so far out of the core, however, they had no significant immediate effect on reactivity, which continued to rise.
9. Coolant bulk boiling led to a rapid rise in power, which Doppler broadening could not counteract.

10. The power rose to 530 MW(th) at 01.23.43 h, and thereafter rose exponentially with a period of a fraction of a second. The reactor went prompt critical; the fuel (uranium oxide ceramic) shattered due to the thermal shock of the sudden power rise, and the cladding melted; the white hot fuel fragments came in contact with the cooling water, and there was a steam explosion at 01.23.48 h. This was followed a few seconds later by a hydrogen explosion; the hydrogen was generated by zirconium—water and graphite—water reactions. There was no nuclear explosion, for reasons stated earlier. The explosions ruptured the containment.

11. The hot graphite moderator caught fire upon exposure to air.

12. Truly heroic efforts by firemen, helicopter pilots and engineers led to the fire being extinguished, and the radioactive release being stopped. by 5 May, ten days after the accident began. In the interim, it is estimated that 20 per cent of the iodine inventory and 12 per cent of the caesium inventory was released to the atmosphere, together with practically all of the noble gas fission products.

Many significant conclusions may be drawn from the accident. However, few, if any, of the following conclusions are likely to affect Western reactor design.

First and foremost, reactors should have no fast-acting positive power coefficients under normal or foreseeable operating conditions. The RBMK design is unique in this respect among all of the world's commercial reactor designs. Secondly, reactor protective systems should shut down the reactor automatically when important safety parameters exceed normal margins. It is insufficient to leave such action to the operators. The protective system should not be easily overridden. Thirdly, reactor shutdown systems should be fast-acting and highly reliable. (This is a 'belt-and-braces' approach: there is no real need for a fast-acting shutdown system if fast-acting reactivity coefficients are negative.) To obtain a sufficient degree of reliability, it may be necessary to have an alternative (secondary) shutdown system. This is common practice with Western designs. Finally, water-cooled reactor designs need sound containment buildings for the reasons previously stated (section 11.3).

The consequences of the accident have been that 31 people died either at the time of the accident or from radiation sickness. In addition, 135 000 people living within 30 km of the plant had to be evacuated from their homes, because of the doses they received. It is possible that 200 additional cancer deaths may occur amongst them. Soviet data suggest that some 10 per cent of the population dies from cancer in any case. Hence, this excess mortality may just be observable with statistical confidence, when the results of epidemiological surveys become known in years to come.

Those further from the plant did not receive significant doses of radiation; the doses measured are of the order of a few millisieverts. If the (pessimistic) linear dose—risk hypothesis is assumed valid at such small doses, then it is possible that several thousand additional cancer deaths may arise in Eastern Europe over the next few decades. However, the natural incidence of the disease will mean that this additional mortality, if it occurs at all, will probably not be observable. In other words, there is no significant increase in the individual risk of death due to cancer as a result of the accident, except among those living close to Chernobyl.

The Chernobyl accident is about as bad a reactor accident as can be envisaged. The reactor core burned without any containment for ten days; it is difficult to postulate an accident which could lead to a greater release of radioactivity to the environment.

11.5 Comparative risks and costs of power generation

No discussion of nuclear reactor safety should be complete without a brief mention of the *relative* risks and costs with respect to other means of generating electricity.

Coal combustion can be quite damaging to the environment; the great London smog of 1952 caused at least 4000 extra deaths due to bronchitis, and led to legislation intended to reduce smoke discharges in the UK. Also, the possibility exists that sulphur dioxide and nitrous oxide emissions from power stations may be causing 'acid rain', leading to damage to forests and lakes; other mechanisms have also been proposed and no clear causal link has yet been established. Carbon dioxide from fossil fuel combustion may lead to a rise in global temperatures via the 'greenhouse effect', though this too, is not clearly established. Furthermore, fossil fuel combustion produces a carcinogenic material, benzo-a-pyrene, which may be causing thousands of deaths each year around the world. (This is calculated by assuming, as for radiation, that there is a linear relationship between dose and risk.) Finally, deep mining of coal leads to a considerable death toll among miners due to accidents and pneumoconiosis.

Hydro-electric power presents hazards due to the possibility of dam failure. The greatest industrial accident in history has been the Gujarati dam failure in India in 1979, which may have caused up to 15 000 deaths. Dam failures in France in 1959 and Italy in 1963 caused 400 and 1800 deaths respectively, and a failure in Pennsylvania in 1889 caused 2200 deaths.

Conservation, too, is not without its hazards. Draught-proofing of buildings can lead to a considerable increase in the concentrations of radioactive radon gas present; radon is released from brick and stone as the daughter product of the decay of trace quantities of natural uranium that is present in the building materials or in the foundations or ground under the building. Using the linear dose—risk hypothesis for the effects of low-level radiation, one may calculate that a significant number of cancer deaths may arise through this exposure route.

No discussion of risk is proper without mention also of cost. Some recently published data for both risks and costs are included in Table 11.1. It should be noted that data like these are among the most contentious in the ongoing public debate about nuclear power. Operating costs for nuclear power stations can show considerable variation, even between stations of a similar type. Likewise, risk data are sensitive to the assumptions made in their calculation.

Table 11.1. Relative risks and costs of power generation

Method of generation	Hazard or other difficulty	Deaths/ GW(e)—year	Relative cost/kWh (Nuclear = 1·0)
Nuclear	Routine discharges	c. 0·05	1·0 *
	Reactor accidents (Chernobyl)	c. 1·0	
	Uranium miners	c. 0·1	
Coal	Cancers − general public	c. 10	1·2
	Miners − accidents	c. 2	
	Miners − pneumoconiosis	c. 8	
	Acid rain	?	
	Greenhouse effect	?	
Draught- proofing	Natural radon from building fabric	c.100 (deaths/ GW(e)−year saved)	Variable
Wind	Inherently unreliable	—	c.1·4
Geothermal	Limited potential	—	c.2·25
Hydro	Dam failure	c. 10	Depends on site
Tidal	Limited potential	—	c.2·5
Photovoltaic	Inherently unreliable	—	c.5
Wave	Hazard to shipping	?	c.6
	Maintenance accidents	?	

*Includes capital, operation, fuel cycle and decommissioning costs

Sources: Central Electricity Generating Board
    ETSU−R30 (1985)
    JH Fremlin, 'Power production − What are the Risks?', Adam Hilger 1985

11.6 Designing for safety

No system can ever be absolutely safe. In a nuclear power station, it is always possible to conceive a scenario whereby a sequence of events could lead to a release of fission products. The best that can realistically be done is to design a system such that the probability of large-scale, uncontrolled fission product release is made negligibly small.

Using the techniques of reliability analysis, including fault tree and event tree analysis, it is possible to estimate the probability (or frequency per annum) of a particular failure sequence occurring. If all failure routes are taken into account, it is then possible to estimate the probability that a given magnitude of fission product release will occur from a nuclear plant. This type of analysis is known as *probabilistic risk assessment.*

It is apparent, therefore, that the degree of safety of a nuclear installation becomes a matter of design choice. Any design can be made safer by incorporating, for example, more diverse shutdown systems or more containment barriers. The question is this: when is a design safe enough?

One suggestion for answering this difficult question has come from F.R. Farmer, who proposed, in 1967, the safety criterion illustrated in Figure 11.9. The criterion is based on the frequency per reactor-year of a given magnitude of release of iodine-131. (As has been shown in Chapter 9, ^{131}I is one of the most toxic of the fission products. It is also one of the most volatile, and is therefore most likely to be released in any reactor accident.) The line on Figure 11.9 thus represents one judgement as to where the dividing line between 'insufficiently safe' and 'safe enough' should lie.

In designing nuclear power stations, it is common practice to confirm, using probabilistic risk assessment, that the frequency of any given

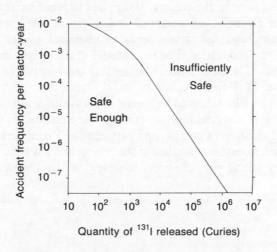

Figure 11.9. The Farmer criterion for reactor safety

magnitude of accident lies within the 'safe enough' region of a safety criterion such as that shown in Figure 11.9. The exact choice of safety criterion is, in the final instance, a matter for political decision.

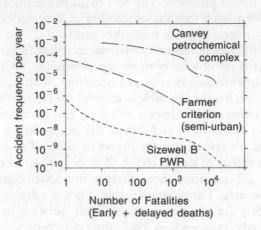

Figure 11.10.   A comparison between the estimated risks of the Canvey petrochemical complex, the Sizewell B PWR power station, and the Farmer criterion for reactor safety
Sources: Health and Safety Executive, 'Canvey, a Second Report', HMSO 1981; Bell G.P., in 'Nuclear Reactor Safety', Ed. F.R. Farmer, Academic Press, New York, 1977; Roberts L.E.J., 'Nuclear Power and Public Responsibility', Cambridge University Press 1984

Probabilistic risk assessment is realistic in that it assumes that accidents can happen. The same technique may be applied to other hazardous industries; dams, for example, are observed to fail with a frequency of about 10^{-4} per annum, and risk assessments of chemical plants have been published in several countries. The magnitudes of these risks may be compared with those of nuclear power stations, by converting the horizontal axis on Figure 11.9 from 'quantity of ^{131}I released' to 'number of fatalities'. (This may be achieved by means of plume dispersal calculations, together with the sort of health physics calculation described in Chapter 9.) Thus Figure 11.10 shows a comparison between the Farmer criterion, and probabilistic risk assessment studies of the Sizewell B PWR power station and the Canvey Island petrochemical complex on the Thames estuary.

Chapter 12

Nuclear fuel processing

12.1 The nuclear fuel cycle

The way in which uranium is used and possibly re-used in nuclear reactors involves a very much more complex series of processes than the use of coal or oil in conventional power stations, in which the fuel goes more or less directly from the mine or oil-well to the power station, is burned and goes up the chimney in smoke to disperse in the atmosphere.

The nuclear fuel cycle can be visualized in Figures 12.1 and 12.2 which show its principal components. The following stages are involved:

1. The mining of uranium-bearing rock, followed by crushing, separation and processing to produce the uranium oxide U_3O_8, known as yellowcake.
2. Transportation of yellowcake to the uranium fuel element manufacturing plant.
3. Chemical, processing of yellowcake, enrichment if required and manufacture of fuel elements.
4. Transportation of fuel elements to the nuclear power station.
5. Storage and use of fuel in reactors (typical in-core time, $1-3$ years) and after unloading, storage of the irradiated fuel for several months.
6. Transportation of irradiated, highly radioactive fuel from the power station to the reprocessing plant.
7. Reprocessing and separation of the irradiated fuel into its three components—depleted uranium, plutonium and fission products.
8. Storage of fission products in liquid form at the reprocessing plant for several years, followed by further processing to produce material in a form suitable for ultimate long-term disposal.
9. Storage of depleted uranium and plutonium, and recycle of these materials to the fuel element manufacturing plant for re-use in reactors.

Alternatively, items 7, 8 and 9 may be replaced by:

10. Storage and ultimate disposal of irradiated fuel without reprocessing.

At each stage of the nuclear fuel cycle there is some element of risk, often no greater than the risks involved in comparable industrial processes

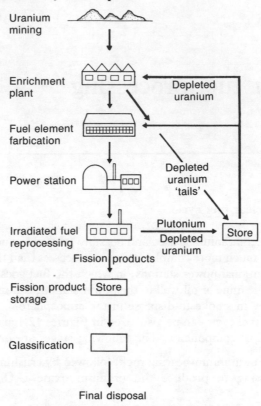

Uranium mining

Enrichment plant

Depleted uranium

Fuel element farbication

Power station

Depleted uranium 'tails'

Irradiated fuel reprocessing

Plutonium

Depleted uranium

Store

Fission products

Fission product storage

Store

Glassification

Final disposal

Figure 12.1. The nuclear fuel cycle

which are accepted without qualms. In the rest of this chapter the several stages of the nuclear fuel cycle will be dealt with more fully and the hazards inherent in each will be described.

12.2 Uranium mining

Uranium occurs quite widely on the earth's surface in small deposits. In the 'western world' the principal uranium resources are located in the U.S.A., Canada, Australia, South Africa, Namibia, Niger, Gabon and France. The mining of uranium might be thought to present a hazard due to the radioactive nature of the material being mined, although uranium-bearing rocks contain only a very small percentage of uranium itself. The radioactivity hazard is only significant in underground mining with inadequate ventilation, for in those circumstances the radioactive gas radon 222 (which is a daughter product of uranium 238) may build up to sufficiently high concentrations as to present a health hazard when inhaled by workers underground. Other radioactive daughter products such as polonium 218 are hazardous if they become attached to dust

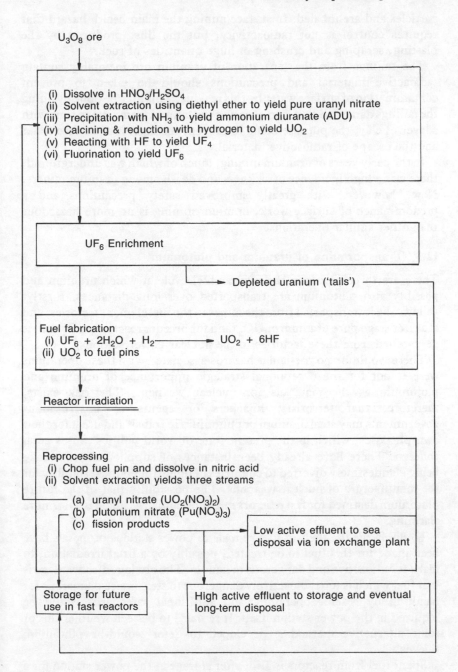

U₃O₈ ore

(i) Dissolve in HNO_3/H_2SO_4
(ii) Solvent extraction using diethyl ether to yield pure uranyl nitrate
(iii) Precipitation with NH_3 to yield ammonium diuranate (ADU)
(iv) Calcining & reduction with hydrogen to yield UO_2
(v) Reacting with HF to yield UF_4
(vi) Fluorination to yield UF_6

UF_6 Enrichment

Depleted uranium ('tails')

Fuel fabrication
(i) $UF_6 + 2H_2O + H_2 \longrightarrow UO_2 + 6HF$
(ii) UO_2 to fuel pins

Reactor irradiation

Reprocessing
(i) Chop fuel pin and dissolve in nitric acid
(ii) Solvent extraction yields three streams
(a) uranyl nitrate ($UO_2(NO_3)_2$)
(b) plutonium nitrate ($Pu(NO_3)_3$)
(c) fission products

Low active effluent to sea disposal via ion exchange plant

Storage for future use in fast reactors

High active effluent to storage and eventual long-term disposal

Figure 12.2. The nuclear fuel cycle

particles and are inhaled. In surface mining the main health hazard that requires control is not radioactivity, but the dust produced by the blasting, scraping and crushing of huge quantities of rock.

The tailings from the separation of uranium ore inevitably contain radioactive material and precautions should be taken to prevent contamination of surface water by this residue. This may be done by lining the tailings dam with clay, and covering the tailings from time to time with a layer of clay, the purpose being to prevent the ingress of surface water and the escape of radioactive material.

In the early years of uranium mining, much of which was underground, there was a high incidence of sickness and death among uranium miners. Now, however, with greatly improved safety precautions and a predominance of surface work, uranium mining is no more hazardous than other similar operations.

12.3 Transportation of uranium and plutonium

There are three points in the nuclear fuel cycle at which uranium and possibly also plutonium are transported over long distances. Firstly, yellowcake is transported from the mines to the fuel element factories, and at a later stage pure uranium or UO_2 (and for breeder reactors PuO_2 also) is despatched from these factories to the nuclear power stations.

There would be no particular hazards associated with these operations were it not for the exceptional strategic importance of uranium and plutonium as the materials for nuclear weapons. The risk exists, therefore, that terrorists, hijackers or agents of unscrupulous governments may steal uranium or plutonium in transit and use it for their own purposes, which in the present state of world politics causes great concern. There have already been instances of supplies of yellowcake being clandestinely diverted to countries of doubtful political stability for the manufacture of nuclear weapons. The possibility that in the future plutonium destined for fast reactors might be similarly stolen is even more alarming.

To discharge the theft of fuel *en route* to power stations proposals have been made for this fuel to be treated, possibly by a brief irradiation, to make it radioactive and dangerous to handle. The obvious disadvantage is that fuel which is too radioactively dangerous to steal is also dangerous for handling by reactor operators, and additional safeguards would be required in the power station itself. It remains to be seen whether this or some alternative method is developed to deter would-be plutonium thieves.

Spent fuel from reactors is also, after storage at the power station for a few months, transported to the fuel processing plant. At this stage the hazard is not that the fuel will be stolen (for it is highly radioactive and far too dangerous for even the most dedicated terrorist to risk handling) but that some accident in transit will allow radioactive material to escape into the environment. At present in the United Kingdom spent fuel from nuclear power stations is transported regularly by road and rail to Sellafield for reprocessing. The lead-lined steel containers in which the fuel is transported have been designed to withstand every conceivable accident and have been subjected to the most stringent tests involving fire, collisions and drop-tests. It is difficult to imagine any accident which could allow the radioactive contents of these containers to be released.

12.4 Chemical processing of yellowcake

Yellowcake is purified uranium ore, U_3O_8. For reactors employing enriched uranium dioxide (UO_2) fuel — the vast majority of power reactors around the world — the next stage is to convert U_3O_8 to uranium hexafluoride, UF_6. UF_6 is a gaseous compound which may be used in diffusion or centrifuge enrichment plants (see below).

The conversion of U_3O_8 to UF_6 takes several steps. Although different processes exist, a typical operation might proceed as follows:

(a) Dissolve yellowcake in a mixture of nitric and sulphuric acids.
(b) By means of *solvent extraction* with diethyl ether, pure uranyl nitrate $UO_2(NO_3)_2$ can be produced in aqueous solution.
(c) The addition of ammonia yields a precipitate of ammonium diuranate (usually referred to as ADU).
(d) The ADU is then calcined (heated) and reduced with hydrogen to yield UO_2.
(e) The UO_2 is then converted to UF_4 by reacting with hydrofluoric acid (HF).
(f) Finally, the UF_4 is converted to UF_6 by reacting it with fluorine.

After enrichment, the uranium hexafluoride is converted back to uranium dioxide by reacting with water and hydrogen. The UO_2 can then be compressed into pellets and assembled into fuel pins. Fuel pins are then arrayed together to yield fuel sub-assemblies (Figure 7.9).

12.5 Enrichment of uranium

In Chapter 3, it was pointed out that for certain types of reactor enriched uranium is necessary to achieve criticality. The most important example is the pressurized water reactor which requires slightly enriched uranium with 2 to 3 per cent ^{235}U. Some current British graphite moderator reactors, e.g. the advanced gas-cooled reactor, use fuel of a similar composition.

The process of enriching uranium involves a partial separation of the ^{235}U and ^{238}U so that the product has a higher concentration of ^{235}U than does natural uranium. The waste, known as the tails, has a lower concentration. Two processes are available for uranium enrichment on a commercial scale. In both of them, natural uranium is converted to the gaseous compound uranium hexafluoride UF_6 and the two isotopes of uranium produce two gases of slightly different density, the $^{238}UF_6$ being slightly more dense than the $^{235}UF_6$. Both processes make use of this slight difference in density to achieve separation of the isotopes.

In the gaseous diffusion process (which was originally developed in America in the 1940s and was the main process of enrichment until the 1970s) uranium hexafluoride is caused to diffuse through a series of porous membranes by virtue of a pressure difference across the membranes. The less dense $^{235}UF_6$ diffuses slightly more rapidly than the denser $^{238}UF_6$, and there is thus some enrichment of the gas which passes through the membrane. The amount of enrichment after passing through a single membrane is very small, and to obtain a significant enrichment many stages are required. For example, to produce $2 \cdot 5$ per cent ^{235}U, about four hundred stages would be needed. Figure 12.3 shows diagrammatically four stages of a gaseous diffusion plant with the feed, enriched and depleted streams indicated.

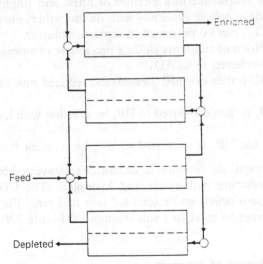

Figure 12.3.   Four stages of a gaseous diffusion plant

The centrifuge process, which has more recently been developed on a commercial scale, involves the use of high speed centrifuges to separate the $^{235}UF_6$ from the $^{238}UF_6$. Uranium hexafluoride is fed into the centrifuge, and at the very high rotational speeds of these machines the $^{235}UF_6$ molecules have a higher concentration near the axis of the centrifuge than in the outer region, and the $^{238}UF_6$ molecules have a higher concentration in the outer region than near the axis. Gas withdrawn from the centrifuge near the axis is thus enriched in ^{235}U, and gas withdrawn near the perimeter is depleted in ^{235}U. Once again, many stages of centrifuging are required to achieve significant enrichments.

Both these processes require large amounts of energy, in one to pump the gas through the many stages of membranes, and in the other to drive the centrifuges. The energy requirement for the centrifuge process is about one-tenth of that required for the diffusion process for the same product, and this leads to a lower cost of enriched uranium from the centrifuge process, which has now largely superseded the gaseous diffusion method of enrichment.

A third process for the enrichment of uranium involves the uses of lasers to selectively excite ^{235}U atoms (in the vapour state) to ionized states at which they can be collected in an electric field. This atomic vapour laser isotope separation (AVLIS) process is being developed in the United States, and a similar method is being studied in Britain. However, the AVLIS process has not yet been developed to a commercial scale, and such is the over-capacity of enrichment facilities in the United States and Western Europe at present that it may be several years before commercial enrichment by the AVLIS process begins. It is expected that it will be significantly cheaper than the centrifuge process.

The cost of enriched uranium depends on two factors, the cost of the natural uranium feed and the cost of enrichment. The amount of natural uranium required to produce 1 kg of enriched uranium depends on the enrichment required and the ^{235}U concentration in the tails. Figure 12.4 shows the flows of uranium, the feed F, product P and tails T, and the fractions of ^{235}U in each of these streams X_f, X_p and X_t.

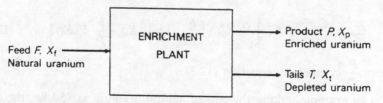

Figure 12.4.   Uranium flows in an enrichment plant

Equating the inflow and outflow of uranium:
$$F = P + T \tag{12.1}$$
and equating the inflow and outflow of ^{235}U:
$$X_f F = X_p P + X_t T \tag{12.2}$$
Combining these equations, it is found that the amount of uranium feed per unit of enriched uranium product is given by:

$$\frac{F}{P} = \frac{X_p - X_t}{X_f - X_t} \qquad (12.3)$$

The influence of the tails concentration X_t is important. For any given value of X_p, the feed required per unit produced increases as X_t increases, and vice versa.

The cost and energy requirement for enrichment are expressed in terms of the number of separative work units (SWU) for a specified enrichment and tails assay. The separative work requirement of enriched uranium is given by the equation:

Separative work requirement $= PV(X_p) - TV(X_t) - FV(X_f)$ (12.4)

where $V(X_p)$ is the value function of the product, $V(X_t)$ is the value function of the tails and $V(X_f)$ is the value function of the feed, natural uranium. The value function of eeach stream is related to the fraction X of ^{235}U in that stream by:

$$V(X) = (1 - 2X)\log_e \frac{1 - X}{X} \qquad (12.5)$$

Although the separative work requirement has units of mass, it can also be expressed either as a cost or as an energy requirement, provided the cost or energy requirement per separative work unit is known. A recent (1986) assessment for the US Department of Energy quoted $70 per kg SWU (£40 per kg SWU) for the centrifuge process.

From the preceding equations it is possible to calculate the cost per kg of enriched uranium, given the cost of natural uranium, the enrichment required, the tails assay and the cost of a separative work unit. Thus if C_u is the cost per kg of natural uranium and C_s is the cost per kg SWU then the cost of 1 kg of enriched uranium is:

$$C_p = \frac{C_u F}{P} + C_s \left\{ V(X_p) + \left(\frac{F}{P} - 1\right) V(X_t) - \frac{F}{P} V(X_f) \right\} \qquad (12.6)$$

where $\dfrac{F}{P}$ is given by equation (12.3).

As an example, taking the cost of natural uranium as £50/kg, the cost of enrichment as £40/kg SWU, the tails assay as $0 \cdot 25$ per cent and the required enrichment as $2 \cdot 5$ per cent, the feed to product ratio $\dfrac{F}{P} = 4 \cdot 84$. The values of V are:

$V(X_p) = V(0 \cdot 025) = 3 \cdot 48$; $V(X_f) = V(0 \cdot 007\ 15) = 4 \cdot 86$;
$V(X_t) = V(0 \cdot 0025) = 5 \cdot 96$;

and the cost of enriched uranium is:

$C_p = (50 \times 4 \cdot 84) + 40\ [3 \cdot 48 + (3 \cdot 84 \times 5 \cdot 96) - (4 \cdot 84 \times 4 \cdot 86)]$
    $= £361$ per kg.

12.6 Fuel reprocessing

After irradiation, fuel may be reprocessed to recover and separate the uranium, the plutonium, and the fission products. The uranium will be somewhat depleted in ^{235}U, but may be mixed with enriched uranium and re-used as reactor fuel. The plutonium may be stored for eventual use in fast reactors. The fission products are stored pending eventual final disposal.

Reprocessing spent fuel begins with the fuel elements being chopped up and dissolved in a strong aqueous solution of nitric acid. This solution is then processed using *solvent extraction*. In this process, an aqueous solution is brought into contact with an organic solvent which, although immiscible with water, has a strong affinity for some of the salts in the aqueous solution. These salts then transfer into the organic solvent.

For the reprocessing of spent nuclear fuel, it was discovered that the organic solvent tri-butyl phosphate (TBP) has an affinity for U^{6+} and Pu^{4+} ions in *strong* acid aqueous solution. Furthermore, the process is reversible since U^{6+} and Pu^{4+} will return to the aqueous phase if that phase is a *weak* acid solution. Finally, the affinity of TBP for plutonium is lessened if the plutonium is in the form of Pu^{3+} instead of Pu^{4+}.

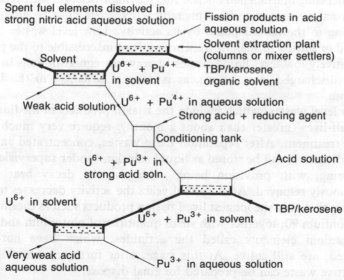

Figure 12.5. A typical flowsheet diagram for a plant to separate the uranium, plutonium and fission products in spent fuel, using the Purex solvent extraction process

By means of these processes, it is therefore possible to, first, separate the uranium and the plutonium from the fission products and second, to separate the uranium from the plutonium. The organic solvent used is generally a solution of TBP in kerosene. A typical flowsheet for such a process is illustrated schematically in Figure 12.5. This process is called the Purex process.

In industrial solvent extraction plant, the aqueous and organic phases are brought into contact in one of two alternative designs: mixer settlers or columns. In mixer settlers, the two phases are (as the name suggests) alternately mixed vigorously, and then allowed to settle. A solvent extraction plant based on mixer-settlers employs a series of such units, with the organic and aqueous phases flowing in opposite directions. In columns, the lighter organic phase is pumped into the bottom of the column while the heavier aqueous phase is pumped in at the top; the column is filled with Raschig rings or similar objects to ensure that the two phases mix thoroughly. In both mixer-settlers and columns, counter current flow is employed to minimize the volumes of liquids required.

12.7 Radioactive waste disposal

The processing of irradiated nuclear fuel inevitably leads to the production of radioactive waste material. This radioactive waste must be disposed of according to the hazard posed by its activity. Low level wastes can be dumped or buried in shallow trenches on land inaccessible to the public. Alternatively low level liquid wastes in very low concentrations in water can be discharged into the sea, as is done at Sellafield in the United Kingdom.

High level wastes, which include the fission products of medium and long half-lives (greater than about a month), require very much more careful treatment. After separation these wastes, concentrated and still very radioactive, can be stored as liquid solutions under supervision and monitoring, with provision being made for the decay heat to be continuously removed. After several years the activity decreases to such an extent that only the longest lived fission products such as caesium 137 and strontium 90, together with small quantities of plutonium and other transuranium elements called the actinides, which have not been separated, are still active. At this stage, after further processing, the radioactive waste can be prepared for final disposal.

At present the intention in several countries including the U.K. and France is to solidify the waste and incorporate it into glass of a suitable composition. Considerable research is being carried out into the properties of suitable glasses; for example, leaching rates in water at the elevated temperatures likely to be encountered in storage sites must be so low that there is no possibility of the glass being dissolved by water within a few thousand years. The glassified waste, probably in the form of cylindrical blocks about 2 m long and 0·5 m diameter, will be enclosed in stainless steel clad with some highly corrosion-resistant metal such as lead, copper or titanium (see Figure 12.6).

These blocks may be stored in suitable repositories under supervision for many years and ultimately may be disposed of either in stable geological formations a few thousand metres underground, or on or under the bed of the deep ocean. No final decisions have yet been made regarding the best method for final disposal, and there is no immediate urgency as 10 years or more will elapse before the problem has to be faced. In the meantime research is being carried out in several countries to test geological strata deep underground for their suitability as sites for ultimate waste disposal. It is the aim of such research to ensure that the rocks in which the waste is placed are geologically stable over periods of millions of years and are able to withstand the elevated temperatures that will be generated by decay heat in disposal sites. It is also necessary to ensure that there are no ground water flows through proposed disposal

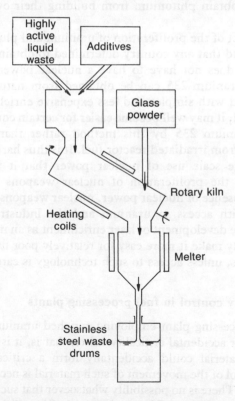

Figure 12.6. The French AVM (Atelier Vitrification Marcoule) process for the glassification of highly active liquid waste

sites that might, if leaching of the glassified waste proceeds too far, carry radioactive materials to the surface of the earth where they might contaminate drinking water supplies.

There is some discussion in the 'western world' at present as to whether or not irradiated fuel should be reprocessed at all. One point of view, advocated in the U.S., is that so long as irradiated fuel remains unprocessed, the plutonium in it remains inaccessible to countries which do not have a reprocessing capability. Thus the danger of widespread proliferation of plutonium for weapons use is reduced. This scheme has certain drawbacks of which the principal one is that by not reprocessing, plutonium and depleted uranium are denied for possible recycling. Furthermore the plutonium is not rendered permanently inaccessible because after a few hundred years the fission product radioactivity will have decayed to very low levels and extraction of the plutonium may become relatively easy. The non-reprocessing option is seen by its advocates as contributing to the non-proliferation of plutonium; however, it does not prevent those countries with nuclear reactors which are determined to obtain plutonium from building their own reprocessing plant.

On the subject of the proliferation of uranium and plutonium, it must be borne in mind that any country determined to obtain weapons-grade fissile material does not have to have a nuclear power programme to achieve this. Uranium 235 can be obtained from natural uranium by enrichment, and with simpler and less expensive enrichment processes being developed, it may well become easier for certain countries to obtain supplies of uranium 235 by this method rather than by separating plutonium 239 from irradiated reactor fuel. It is thus hardly justifiable to say of the large scale use of nuclear power that it will be entirely responsible for the proliferation of nuclear weapons throughout the world. In the absence of nuclear power, nuclear weapons can be made by any country with access to uranium and the industrial capacity for enrichment. The development of laser enrichment as an industrial process would potentially make it more easy for relatively poor nations to possess nuclear weapons, unless access to such technology is carefully regulated.

12.8 Criticality control in fuel processing plants

In any fuel processing plant employing enriched uranium or plutonium, the potential for accidental criticality exists; that is, it is conceivable that some fissile material could accidentally form a critical configuration. Stringent control of the movement of such material is necessary to prevent such accidents. There is no possibility whatsoever that such accidents could lead to nuclear explosions. However, a criticality will lead to a burst of gamma rays and neutrons which will give any plant operators standing nearby – unless they are behind shielding – fatal doses of radiation.

Several accidental criticalities have occurred around the world, and in some cases deaths due to high radiation doses causing irreparable damage to the central nervous system have resulted. It is for this reason that strict controls are put on the quantity of fissile material that may be in any given plant area at any time. Similarly, care must be taken over the design of plant which contains fissile material in solution. For example, the possibility of the accidental addition of a chemical, which might cause the fissile material to be precipitated out of solution, must be considered; such an event could conceivably lead to the precipitate forming a critical mass at the bottom of a storage tank.

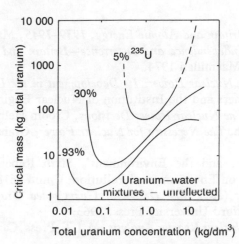

Figure 12.7. Minimum critical masses of uranium at different enrichments. The lines correspond to the masses of uranium for which k_{eff} equals one, i.e. just critical

For the initial design of fuel processing plant, and in particular for plant elements employing simple geometries, it is possible to use diagrams such as Figure 12.7. It is apparent that the potential for accidental criticality is greatly increased for highly enriched uranium. For detailed design, it is necessary to employ numerical techniques for assessing the effective neutron multiplication factor k_{eff}. By such means, it is possible to design plant so that k_{eff} is very much less than one for all foreseeable plant states, thereby ensuring that accidental criticality does not occur (see Appendix 3.3).

A list of books for further reading

M. Gowing, *Britain and Atomic Energy, 1939–1945*, Macmillan, 1964.

M. Gowing, *Independence and Deterrence—Britain and Atomic Energy, 1945–1952*, Macmillan 1974.

R. F. Pocock, *Nuclear Power: Its Development in the United Kingdom*, Unwin Brothers and the Institution of Nuclear Engineers, 1977.

R. Williams, *The Nuclear Power Decisions*, Croom Helm Ltd., 1980.

G. Greenhalgh, *The Necessity for Nuclear Power*, Graham & Trotman, 1980.

'Nuclear Power and the Environment', Sixth Report of the Royal Commission on Environmental Pollution, Cmnd 6618, 1976.

D. R. Poulter (ed.), *The Design of Gas Cooled Graphite Moderated Reactors*, Oxford University Press, 1963.

A. F. Henry, *Nuclear Reactor Analysis*, M.I.T. Press, Cambridge, Mass., 1975.

A. M. Judd, *Fast Breeder Reactors*, Pergamon Press, 1980.

J. R. Lamarsh, *Nuclear Reactor Theory*, Addison-Wesley, 1966.

J. Lewins, *Nuclear Reactor Kinetics and Control*, Pergamon Press, 1978.

R. L. Murray, *Nuclear Energy* (2nd Edition), Pergamon Press, 1980.

T. J. Connolly, *Fundamentals of Nuclear Engineering*, Wiley, 1978.

S. Glasstone and A. Sesonske, *Nuclear Reactor Engineering*, Van Norstrand Reinhold Company, 1981.

S. Rippon, *Nuclear Energy*, Heinemann, 1984.

'Nuclear Energy: a professional assessment', Watt Committee on Energy, Report No. 13, 1984.

W. Marshall (ed.), *Nuclear Power Technology* (3 vols), Clarendon Press, 1984.

E. E. Pochin, *Nuclear Radiation: Risks and Benefits*, Oxford University Press, 1983.

E. Addinall and H. Ellington, *Nuclear Power in Perspective*, Kogan Page, 1982.

'Chernobyl − A Technical Appraisal', British Nuclear Energy Society, Thomas Telford, 1987.

'Advances in Power Station Construction', Central Electricity Generating Board, Barnwood, Pergamon, 1986.

F. R. Farmer (ed.), *Nuclear Reactor Safety*, Academic Press, 1977.

J. F. Flagg (ed.), *Chemical Processing of Reactor Fuels*, Academic Press, 1961.

J. H. Fremlin, *Power Production − What Are The Risks?*, Oxford University Press, 1987.

N. J. McCormick, *Reliability and Risk Analysis − Methods and Nuclear Power Applications*, Academic Press, 1981.

L. E. J. Roberts, *Nuclear Power and Public Responsibility*, Cambridge University Press, 1984.

J. R. Thomson, *Engineering Safety Assessment − an Introduction*, Longman, 1987.

Constants and conversion factors

Avogadro's number $6\cdot023 \times 10^{26}$ atoms/kilogram atom (molecules/kilogram mole)

Boltzmann's constant $1\cdot38 \times 10^{-23}$ joule/Kelvin

Speed of light $2\cdot998 \times 10^{8}$ metres/second

Proton mass $1\cdot007\ 277$ unified atomic mass units

Neutron mass $1\cdot008\ 665$ unified atomic mass units

Electron mass $0\cdot000\ 549$ unified atomic mass units

Electron charge $4\cdot8 \times 10^{-10}$ electrostatic unit
      or $1\cdot602 \times 10^{-19}$ coulomb

Energy: $1\ \text{MeV} = 10^{6}\ \text{eV} = 1\cdot602 \times 10^{-13}\ \text{J} = 1\cdot602 \times 10^{-6}\ \text{erg}$

Mass: $1\ \text{u} = 1\cdot6604 \times 10^{-27}\ \text{kg}$

Mass-Energy: $1\ \text{u} = 931\ \text{MeV}$

Power: $1\ \text{W} = 1\ \text{J/s}$

Burnup: $1\ \text{MWd} = 8\cdot64 \times 10^{10}\ \text{J}$

Cross-section: $1\ \text{barn} = 10^{-24}\ \text{cm}^2$

Radioactivity: 1 becquerel = 1 disintegration/second
       1 curie = $3\cdot7 \times 10^{10}$ disintegrations/second

Radiation absorbed dose: 1 gray = 1 J/kg

Radiation dose equivalent: 1 sievert = 1 J/kg

Cross-sections and other data for materials used in nuclear engineering

Fuels

	Density g/cm³	σ_o barns	σ_f barns	σ_s barns	ν	η
Natural uranium	18·9	3·4	4·2	8·3	—	1·32
^{235}U	18·7	101	579	10	2·42	2·07
^{238}U	18·9	2·72	0	8·3	—	—
^{239}Pu	19·6	266	742	9·6	2·93	2·12

Moderators

		Density g/cm³	σ_o barns	σ_s barns	ξ	$\bar{D}$ cm	L^2 cm²	$L_s^2(\tau)$ cm²
Water	H_2O	1·0	0·66	~50	0·920	0·16	8·1	27
Heavy water	D_2O	1·1	0·001	10·6	0·509	0·87	30 000	131
Graphite	C	1·6	0·0045	4·7	0·158	0·84	2650	368
Beryllium	Be	1·85	0·0092	6·1	0·209	0·50	480	102

Structural, control and other materials

	Density g/cm³	σ_o barns	σ_s barns
Boron	2·3	759	4
Nitrogen	gas	1·85	10
Oxygen	gas	0·0002	3·8
Sodium	0·97	0·53	4
Magnesium	1·74	0·063	4
Aluminium	2·7	0·232	1·4
Sulphur	2·07	0·52	1·1
Iron	7·87	2·56	11
Zirconium	6·8	0·182	8

Notes: Cross-sections for capture and fission are 2200 m/s values. Scattering cross-sections are values for thermal neutrons and are in general (with the exception of H_2O and D_2O) constant at thermal energy.

The capture cross-sections of graphite and heavy water (and hence also their diffusion lengths) are very sensitive to impurities. The value of σ_o given for graphite is for commercial reactor grade graphite which has minor impurities. The value of σ_o given for D_2O is for pure material.

Computational methods in nuclear engineering

A3.1 Neutron diffusion computer codes

Chapters 4 and 5 presented an analytical treatment of nuclear theory. While understanding such theory provides insight into the physics of nuclear reactor calculations, it is nevertheless of only limited practical value; the nature of the assumptions that have to be made compromise the accuracy that can be achieved. These assumptions include one (or at most two) neutron energy groups and a homogeneous core.

By contrast, however, computer solutions of the steady state neutron diffusion equation (4.34) are possible, in principle, for any number of neutron energy groups. Although two group solutions usually suffice for thermal reactors, many more neutron groups (typically 30) need to be considered for fast reactors; computer solutions then become essential. Furthermore, numerical solutions allow inhomogeneity and anisotropy within the reactor core to be taken into account.

Numerical solutions of the neutron diffusion equation may be achieved by the method of finite differences. For simplicity, this is illustrated below using a one-group, one-dimensional model where, over each mesh length, the lattice parameters D and Σ_A are taken as constant.

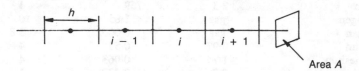

Figure A3.1. One dimensional finite-difference model of the neutron diffusion equation. Homogeneous approximations for D and Σ_a are usually employed, where the fuel element, cladding, coolant and structural material are 'smeared' over each mesh volume

Considering the volume around mesh point i, the rate of neutron absorption is given by $\Sigma_{ai}\phi_i h A$ and the leakage from the volume is given by the following expressions:

$$\text{Leakage to left} \quad = D_L\left(\frac{\phi_{i-1} - \phi_i}{h}\right)A \tag{A3.1}$$

$$\text{Leakage to right} = D_R\left(\frac{\phi_{i+1} - \phi_i}{h}\right)A \tag{A3.2}$$

where D_L and D_R are the appropriate average values for the local neutron diffusion coefficients. These are approximately given by

$$1/D_L = 1/D_i + 1/D_{i-1} \tag{A3.3}$$

$$1/D_R = 1/D_i + 1/D_{i+1} \tag{A3.4}$$

Hence from the neutron diffusion equation (4.34) we obtain

$$D_L\left(\frac{\phi_{i-1} - \phi_i}{h}\right)A + D_R\left(\frac{\phi_{i+1} - \phi_i}{h}\right)A - \Sigma_{ai}\phi_i h A + S_i h A = 0 \tag{A3.5}$$

where S_i is the neutron source per unit volume at mesh point i. From equation (4.48) we get that

$$S_i = k_\infty \Sigma_{ai} \phi_i \tag{A3.6}$$

if the volume at mesh point i contains fissile material; in non-fissile material (e.g. reflector material around the core) S_i will be zero. Hence equation (A3.5) becomes a series of n simultaneous equations where n is the number of mesh points. In matrix notation,

$$\mathbf{F}\phi = 0 \tag{A3.7}$$

where $\mathbf{F}$ is a matrix of coefficients and ϕ is a column vector of neutron fluxes. Equation (A3.7) can be solved by standard numerical techniques.

The above method is adequate provided the absorption is not too large and that we are not too near boundaries. (In these cases some inaccuracies may arise.) The method may be extended to two- or multi-group calculations by including the effects of neutron slowing-down collisions in equation (A3.5), as shown in equations (5.3) and (5.4). The method can also be readily extended to 2 or 3 dimensions by including terms describing leakage in those directions.

A3.2   Reactor kinetics computer codes

Analysis of the time behaviour of a reactor system is important during start-up conditions, power changes and under fault conditions. The simple one-group equations of reactor kinetics as given by equations (8.13) and (8.15) are normally sufficient for analysing transient behaviour of thermal reactors; however we are usually interested in spatial variation of flux (at least in one dimension) as well as variation with time. If we consider leakage in one dimension only, e.g. by considering a fuel element in the centre of a cylindrical reactor where leakage in the radial and tangential directions will (by symmetry) be zero, then equation (8.13) can be modified to yield

$$\frac{1}{v} \frac{\delta\phi}{\delta t} = D \frac{\delta^2\phi}{\delta z^2} + k_{\text{eff}}(\rho - \beta)\Sigma_a\phi + \sum_{i=1}^{6} \lambda_i C_i \qquad \text{(A3.8)}$$

where z is the vertical ordinate, and equation (8.15) yields

$$\frac{\mathrm{d}C_i}{\mathrm{d}t} = k_{\text{eff}}\beta_i\Sigma_a\phi - \lambda_i C_i \quad \text{(six equations)} \qquad \text{(A3.9)}$$

Here $C_i = C_i(z, t)$ and $\phi = \phi(z, t)$, ρ and hence k_{eff} are parameters, and all other terms are coefficients. Equations (A3.8) and (A3.9) may be converted, using finite-difference approximations for the leakage term $D \frac{\delta^2\phi}{\delta z^2}$ as in (A3.8) to yield a matrix equation of the form

$$\mathbf{V}^{-1} \frac{\delta\psi}{\delta t} = \mathbf{F}\psi \qquad \text{(A3.10)}$$

Here $\mathbf{V}$ and $\mathbf{F}$ are coefficient matrices, and ψ is a matrix of the form

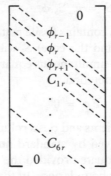

where subscript r refers to the rth out of a total of k mesh points in the z direction. Equation (A3.10) can be further arranged in this form

$$\mathbf{V}^{-1}\Delta t^{-1}(\psi_n - \psi_{n-1}) = \mathbf{F}\psi_n \qquad \text{(A3.11)}$$

where subscript n refers to time variation (Figure A3.2), and Δt is the time step size.

This set of simultaneous equations can be solved using standard techniques to yield solutions for the spatial and time variation of reactor flux for a constant value (or at most a predetermined variation) of k_{eff}. However in reactor fault studies, significant changes in fuel and coolant temperatures may occur. In those circumstances temperature feedback must be taken into account. To do this, the thermal−hydraulic equations governing the relationships between coolant flow, fuel temperature and neutron flux (as introduced in Chapter 6) must be considered, in order that the variation of temperature as a function of time and space, $T(z, t)$ can be established. The effect of such temperature variation upon reactivity may be determined

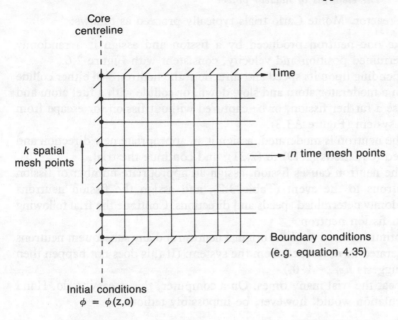

Figure A3.2. Finite-difference scheme for kinetic calculation

from previously-assessed values for the temperature coefficient of reactivity, α (equation (8.34),

$$\alpha = \frac{\delta\rho}{\delta T}$$

It is thus apparent that the neutron flux affects the temperature, which affects the reactivity, which affects the neutron flux (see Figure 8.6). We say that the differential equations governing the neutron flux and the reactor core thermal hydraulics are *coupled*. (For further discussion of methods of solution of these equations, readers are advised to read, for example, **J. Graham**, *Fast Reactor Safety*, Academic Press, 1971.)

A3.3 Monte Carlo computer codes

On some occasions we are only interested in determining a value for k_{eff} for a given system; for example, it might be required to demonstrate that in a storage pond containing irradiated fuel, k_{eff} was very much less than unity for all conceivable fuel configurations (i.e. that a criticality was not possible − see also Chapter 12). The most suitable method of solving such a problem is to use a so-called 'Monte Carlo' code.

The Monte Carlo method goes right back to the statistical nature of the neutron transport problem. The technique consists of tracing in detail the path followed by individual neutrons drawn at random from the population in the reactor. Monte Carlo trials typically proceed as follows:

1. Take one neutron produced by a fission and assign it a randomly determined position and velocity, consistent with Figure 2.6.
2. Depending upon its speed and direction, the neutron will either collide with a moderator atom and slow down, or collide with a fuel atom and cause a further fission, or be captured without fission, or escape from the system (Figure A3.3).
3. If the neutron is moderated, assign it an appropriate new direction and speed (equations (2.19) to (2.21)) and continue the trial.
4. If the neutron causes fission, assign an appropriate number of fission neutrons to the event (Table 3.2), and assign the fission neutrons randomly determined speeds and directions. Continue the trial following each fission neutron.
5. Continue the trial until the original neutron and any subsequent neutrons generated have escaped from the system. (If this does not happen then it suggests $k_{eff} > 1 \cdot 0$.)
6. Repeat the trial many times. On a computer, this is easy to do. Hand calculation would, however, be impossibly tedious.

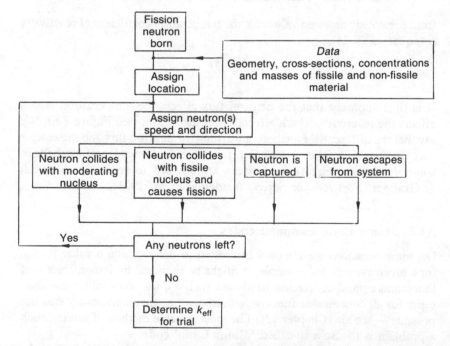

Figure A3.3.   A single Monte Carlo trial for determining k_{eff} for a system which is believed to be sub-critical. Some thousands of such trials are required to yield accurate results

7. The average ratio of (number of first generation neutrons):(number of second generation neutrons) is then equivalent to k_{eff}.

The Monte Carlo method is a statistical method. The results can therefore have confidence limits assigned to their accuracy. The method was developed largely by von Neumann and Ulam during the Second World War, although full development of the technique awaited the advent of large digital computers. The method can deal with any inhomogeneity or anisotropy in a potentially critical array of fissile material.

Nuclear Power Worldwide

The following table and notes summarize the status of nuclear power in those countries with installed nuclear electricity generating capacity in excess of 2000 MWe. Figures relate to 1986 and are based on data from the International Atomic Energy Agency Power Reactor Information System and the Journal of the European Nuclear Society.

Table A4.1. Numbers of power reactors in operation in 1986

Country	Installed capacity MWe	Number of units	Types of reactors		Nuclear electricity supply in 1986	
					TWh	% of total
Belgium	5 600	7	PWR	− 7	37·1	67
Canada	11 250	18	PHWR	− 18	67·2	14·7
Czechoslovakia	2 800	7	PWR	− 7	16·2	21
Finland	2 310	4	PWR	− 2	18·0	38·4
			BWR	− 2		
France	44 690	49	PWR	− 43	241	70
			GCR	− 4		
			FBR	− 2		
Germany (West)	18 950	21	PWR	− 11	120	30
			BWR	− 7		
			HTGR	− 2		
			FBR	− 1		
Japan	25 800	35	PWR	− 16	166	25
			BWR	− 17		
			GCR	− 1		
			LWCHWR	− 1		
Korea (South)	5 380	7	PWR	− 6	26·6	43·6
			PHWR	− 1		
Spain	5 500	8	PWR	− 5	37	29
			BWR	− 2		
			GCR	− 1		
Sweden	9 650	12	PWR	− 3	67	50
			BWR	− 9		
Switzerland	2 950	5	PWR	− 3	21·3	39
			BWR	− 2		
Taiwan	4 900	6	PWR	− 2	25·8	44
			BWR	− 4		

Country	Installed capacity MWe	Number of units	Types of reactors		Nuclear electricity supply in 1986	
					TWh	% of total
United Kingdom	10 000	38	GCR	− 26	52.6	20
			AGR	− 10		
			FBR	− 1		
			SGHWR	− 1		
United States	84 600	99	PWR	− 63	414	16.6
			BWR	− 35		
			HTGR	− 1		
USSR	27 600	50	PWR	− 20	148	10.6
			BWR	− 1		
			LWGR	− 26		
			FBR	− 3		

Abbreviations:
PWR − Pressurized Water Reactor
BWR − Boiling Water Reactor
GCR − Gas Cooled Reactor
AGR − Advanced Gas Cooled Reactor
HTGR − High Temperature Gas Cooled Reactor
FBR − Fast Breeder Reactor
PHWR − Pressurized Heavy Water Reactor (CANDU)
LWCHWR − Light Water Cooled Heavy Water Moderated Reactor
SGHWR − Steam Generating Heavy Water Moderated Reactor
LWGR − Light Water Cooled Graphite Moderated Reactor

Notes

Belgium

Belgium's early interest in nuclear energy was due to the fact that in the 1940s the Belgian Congo (as its African colony was then called) was one of the few countries producing uranium ore. A treaty with the USA gave the latter priority in the purchase of uranium from that source, and led to co-operation between these two countries which resulted in Western Europe's first PWR at Mol in Belgium in 1962. More recently Belgium and France have co-operated in nuclear power station construction and electricity supply.

Canada

By virtue of having some of the world's first heavy water production plant, Canada's nuclear power programme has from its earliest days in the 1940s been almost exclusively based on the natural uranium fuelled, heavy water moderated and cooled reactor of the pressure tube design, the CANDU. All Canadian power reactors are of this type, and are located in the eastern provinces. CANDU reactors have been exported to India and Argentina, and these two countries are now developing their own designs of PHWR.

Czechoslovakia

Apart from Russia among the Comecon countries, Czechoslovakia has the most developed programme of nuclear power, with the intention that all new electricity generating plant in the next two or three decades will be nuclear, leading to 50 per cent generation from this source. The giant Skoda engineering company is the country's principal constructor of nuclear power plant, and it is also engaged in nuclear power plant construction in Poland, East Germany and Hungary. The standard type of reactor in all these countries is the Russian-designed VVER, a version of the PWR.

Finland

Situated between western and eastern spheres of influence, Finnish electricity utilities have commissioned two Russian-designed pressurized water reactors, VVERs, and two Swedish-designed (ASEA–ATOM) boiling water reactors. There are no present plans for new nuclear plant to be ordered.

France

The earliest nuclear reactors built in France were gas-cooled, graphite-moderated reactors similar to British Magnox reactors. Only a few of these were built and in the 1970s France changed to PWRs, at first building American designs under licence and more recently developing its own design of PWR built by the Framatome company, one of the giants of the world's nuclear industry. In 1973, at the time of the first sudden rise in oil prices following the formation of the Organisation of Petroleum Exporting Countries (OPEC), France (having very limited indigenous resources of fossil fuels) embarked on a concerted programme of nuclear power plant construction, concentrating on the PWR and also playing a leading role in FBR development in Western Europe. France is now Western Europe's leader in terms of installed nuclear power capacity, with a national electricity generating system largely dependent on this source. In contrast with several other Western European countries, there is apparently no political or social opposition to the development and use of nuclear power in France.

West Germany

Early development of nuclear power in West Germany was based on both the BWR and PWR types, built under licence. Now nearly all development is of PWRs. In the last several years West German reactors have topped the world list for individual power generation. In 1986 the 1365 MWe Grohnde PWR produced 10·8 TWh. West Germany has also pioneered the development of the high temperature gas-cooled reactor (HTGR) with spherical fuel elements. This reactor was designed to be fuelled with thorium for breeding uranium-233, but to date the reactor has been fuelled with conventional uranium fuel. There is at present considerable political opposition to the development of nuclear power in West Germany, particularly since the Chernobyl accident in 1986, and one major political party is committed to close all nuclear power plant within ten years.

Japan

With virtually no indigenous resources of fossil fuels, Japan is committed to a major programme of nuclear power which is intended to supply about 60 per cent of electricity by 2030. This programme is largely based on PWRs and BWRs, with development of FBRs also.

South Korea

Continued expansion of nuclear power is planned, based on PWRs supplied by American companies with arrangements for technology transfer to enable South Korea to build its own reactors in due course.

Spain

There is at present a moratorium on further nuclear power construction in Spain, and work on four PWRs is halted.

Sweden

The major part of Sweden's nuclear power generation is derived from the indigenous ASEA—ATOM design of BWR which has proved very successful. With practically no fossil fuel resources, Sweden is largely dependent on nuclear power and hydro-electric power for electricity supplies. Strong political and social opposition to nuclear power led to a Parliament decision in 1980 to phase out nuclear power by 2010, the end of the lifetime of the existing reactors. The Chernobyl accident of 1986 served to reinforce this decision. It remains to be seen what energy resources will be able to replace lost nuclear power in Sweden.

Switzerland

Political uncertainty about the future of nuclear power, particularly in the wake of the Chernobyl accident, has increased. However, despite the availability of hydro-electric power in Switzerland, the country has become dependent on nuclear power and imports of French nuclear-generated electricity to maintain adequate supplies.

United Kingdom

The first British power reactors, the four Calder Hall reactors commissioned in 1956—7 and designed primarily for plutonium production for military purposes, were also the world's first nuclear reactors to supply large quantities of electricity for peaceful purposes. The earliest of the commercial gas-cooled reactors built to the same design have now reached their design life of 25 years, and it is likely that they will be licensed to operate for a further 10 years. The more recent advanced gas-cooled reactors have been less successful, having been beset by construction and operating problems and performing to date with disappointingly low load factors. Following a long and detailed public enquiry, the Government has authorized the construction of a PWR at Sizewell, and this is expected to be the first of a series of about six such reactors. At Dounreay the Prototype Fast Reactor has overcome early problems and is operating at its design power. Britain, France and West Germany are co-operating in the development of a commercial European FBR. As in West Germany, nuclear power has become a political issue with some political parties in the UK committed to phasing it out.

United States

For forty years the world leader in nuclear power, and the originator of the pressurized and boiling water reactors, the two most important types in the world at present, built under licence and copied in many other countries. In the 1970s the US was embarked on a large programme of nuclear power plant construction, but the major accident involving a PWR at Three Mile Island in 1979 was a great setback. In the following years nuclear power plant construction was severely curtailed while safety studies were carried out and many electricity utilities reviewed their plans for nuclear power. As a result, the construction of some power plant was stopped, orders cancelled and some plant converted to fossil fuel firing. At the same time, partly as a result of the extra cost of added safety systems, the relative costs of nuclear and coal fired plant have shifted towards cheaper coal fired plant. Expansion of nuclear power is now proceeding much more slowly, with construction of existing plant continuing but no new plant planned.

USSR

Forty years of development of nuclear power in the USSR has been based on two designs — the VVER, a version of the pressurized reactor, and the

RBMK, a light water cooled graphite moderated reactor of a design unique to Russia. In April a criticality accident followed by an explosion and fire in one of the Chernobyl RBMK reactors led to the world's worst nuclear reactor accident, resulting in a massive release of radioactivity into the environment. About 135 000 people were evacuated from an area within 30 km of the power station, and a massive decontamination of the area around the reactor has been carried out. As a consequence of this accident modifications to the RBMK reactors have been carried out to cure the design weaknesses which made the accident possible. Despite this very serious accident there is no halt to the Russian expansion of nuclear power which is intended to provide 30 per cent of electricity generation by 2000.

Index